Deleuze and the Map-Image

Deleuze and the Map-Image

Aesthetics, Information, Code, and Digital Art

Jakub Zdebik

BLOOMSBURY VISUAL ARTS
LONDON • NEW YORK • OXFORD • NEW DELHI • SYDNEY

BLOOMSBURY VISUAL ARTS
Bloomsbury Publishing Inc
50 Bedford Square, London, WC1B 3DP, UK
1385 Broadway, New York, NY 10018, USA
29 Earlsfort Terrace, Dublin 2, Ireland

BLOOMSBURY, BLOOMSBURY VISUAL ARTS and the Diana logo
are trademarks of Bloomsbury Publishing Plc

First published in the United States of America 2019
Paperback edition published by Bloomsbury Visual Arts 2024

Cover design by Maria Rajka
Cover image: *Geological timescale* © Mary Evans / Natural History Museum

A catalog record for this book is available from the Library of Congress.

ISBN: HB: 978-1-5013-4678-1
PB: 978-1-3504-3693-0
ePDF: 978-1-5013-4680-4
eBook: 978-1-5013-4679-8

Typeset by Integra Software Services Pvt. Ltd.

For Eliot and Helena

Contents

List of Figures

Acknowledgments

I would like to thank first and foremost Margaret Michniewicz at Bloomsbury for making this book possible. Thank you to the anonymous peer reviewers of this book for their constructive and helpful suggestions. My gratitude also goes to the Department of Visual Arts at the University of Ottawa for fostering an ambiance conducive to creative and intellectual pursuits. Thank you to Françoise Moreau-Johnson and the members of the writing workshops who, at different times, read through different parts of the book. I am grateful to the AEPI groupe de recherche for permitting me to present my findings at talks and face productive scrutiny. I want to thank, as always, Daniel Vaillancourt, Calin Mihailescu, and Gary Genosko for their constant support. I am very grateful to Kane X. Faucher and Paul Manning at Semiotic Review for their insightful comments while reviewing and subsequently publishing an article they permitted to be republished in these pages. I am very appreciative of the artists and their assistants and, in some cases, their estates, for their participation in this project. To Colette Stoeber, who edited the book, thank you.

I would like to thank my family and friends for their encouragement. I wanted to use this opportunity to remember Trudy, who passed away during the drafting of the book and who is missed. Most of all, I would like to thank Barbara for her support and steadfastness, without whom my academic endeavors would amount to wishful thinking; my deepest gratitude goes to her.

Introduction: Map as Fluctuating Image

The map covers the territory. This is a familiar image for readers of Gilles Deleuze. Jorge Luis Borges writes a short story, a fragment, about how an empire with a propensity for mapmaking—hell-bent on perfecting the craft, each map more detailed than the last—finally devises a map so comprehensive that it covers the entirety of the land. The punch line is already there, mid-story. The map is so large that it matches, point by point, every detail of actual space. The representation has taken over that which it is supposed to represent. Unmentioned, but obviously part of the joke, is that we are left wondering how such a device would work. Are we to imagine that the map covers the entire ground: all architectural elements of the cities and the topographic features of the landscape? We also ponder the elastic conception of how a map, a familiar image, can become abstract and unfamiliar when we zero in on it. Are the subjects of the empire made to live underneath the map? How was it unfolded? Was it ever folded? How was it produced, kept, protected from the weather?

All of these unanswered questions are arresting and create the richness of the story. They also point to essential questions about representation, the function of mapping, and the relationship of terrain to cartography. No wonder this example comes up in several poststructuralist texts: those of Jean-François Lyotard, Jean Baudrillard, Guy Debord.[1] But what happens next in Borges's story is not often discussed. The dénouement, as it were, of the paragraph-long story is about the unraveling of the map. The generations following the time of the cartographers, Borges writes, were not as interested in mapping as their predecessors. They let the map perish through exposure to the elements (which indicates that it was deployed), and its weathered fragments are discarded in one corner of the empire, where they are used by animals and the destitute as a shelter—demonstrating the map's lowly position. And its unraveling initiates a darker thread running through the following pages hinted at more or less directly throughout each chapter and elucidated in the last one, that of a posthuman, agentless, dystopian map.

The map in Borges's story has a wonderful quality. It is amorphous. Its shape is impossible to grasp. The story, a short paragraph, belies the enormity of the object it captures. I will examine this nebulous quality of maps in this introduction by surveying the instances where maps intrude into Deleuze's writing about art, looking principally at the works of Francis Bacon, Alain Resnais, and Franz Kafka, as they are divided along media lines (painting, film, literature). This is to show that spatial conceptualization is an active component of Deleuze's aesthetics in his analysis of art and that the map featured within these analyses of art is never selfsame.

After showing how the map-image drifts and transforms throughout Deleuze's corpus, I will then explain how the concept—or rather, the image—of the map is threaded throughout the chapters of this book (*image* rather than *concept* of a map because the image can service different concepts while always retaining its cartographic characteristics). In the first chapter, I will review the occasions in which Deleuze discusses Johannes Vermeer's use of the map and space and consider how the map becomes an image representing the notion of information according to art historian Leo Steinberg. In the second chapter, referring to Deleuze and Guattari's *A Thousand Plateaus*, I will discuss cartography as a component of the principles making up part of the rhizome. In the third chapter, I will consider the notions of stratification, boundary, and line that come out of the multidisciplinary and intermedial work of Cory Arcangel, a digital artist whose works shuttle between the digital and celluloid film. In the fourth chapter, I will look at notions of virtuality, swarming, and the grid in the works of John F. Simon Jr., an algorithmic artist who creates works that marry conceptualism with notions of geography and urban landscapes. In the fifth chapter, I will explore the image of the island in diverse media, namely, in a work of Internet art, a projected sculpture, and a science fiction film. The image of the map in all of these instances is associated with heterogeneous concepts. And finally, the book's last chapter focuses on examples of the map-image as a way of wading into the posthuman dimension of art, where surveillance, embeddedness of code, network control, and human obsolescence are part of a new type of image. This type is exemplified by the works of Hajra Waheed, which grapple with satellite views and images seen from above, as well as by the operative image captured in its various incarnations by Harun Farocki.

Bacon: Desert, marshland, ocean

The map intrudes into the figure. In his treatment of Bacon's figurative painting, which teeters on abstraction, Deleuze invokes Borges's map—the one, described above, in "On Exactitude in Science," a fragment-length story attributed to a fictional traveler—and compares it to a diagram, a set of traits recomposing themselves in the painting, taking over the painted surface: "The entire painting is diagrammatic,"[2] Deleuze writes, like a signal scrambling transmission. Too many traits and the figure is lost. We end up with a Pollock, or a map that covers a territory. Of course, the map, despite its opening toward multiplicities, is a static, representational snapshot of the dynamic, abstract diagram. And whereas the diagram operates in the interstices between the virtual and the actual, the map is an actual view of the virtual potential: it does not embody but points to and orients through representation. In Borges's story, the map is the same size as the territory it represents, it is a "Map of the Empire whose size was that of the Empire."[3] The abstract painting, by extension, is like that too, like the map "coinciding point for point" with the territory and considered, subsequently, of no use.[4] The information provided by the map is redundant. The map here is the thing itself, the trees as the forest; we are so close to the canvas, it appears blurry. It is easy to understand why a map superimposed on the very land it represents would be of limited use; but what is the information Deleuze wanted to get out of abstract art?

We are made aware that Deleuze does not look too kindly on abstract art. In fact, in the "Percept, Affect, and Concept" chapter of *What Is Philosophy?*, in which Deleuze and Guattari write about art more comprehensively than any other place in their system—enumerating seemingly every major touchstone of Western aesthetics (visual arts, music, and literature)—the standard-bearers of purely abstract painting are scarce. And when they do show up—Yves Klein, for example—they are not displayed in a flattering light.[5] Klein, known among other things for his International Klein Blue, a color he developed and registered, is an artist who practiced at the cusp of conceptual and abstract art. Among his most recognized works are *Leap into the Void* (1960), a black-and-white photograph of the artist freely throwing himself from a cement fence onto a sidewalk as if plunging into an ocean, and *Anthropométrie de l'époque bleue* (1960), in which the imprint of paint-covered bodies on the surface of a white canvas continues the idea of the body against a ground. Here, we might think of Klein's blue monochromes in geographic terms as oceans or skies. But as either abstract or conceptual, this art is informational, or "informative," and its definition modulates between art or non-art according to the point of view of the spectator, in a sort of binary code of appreciation.[6] Between "sensation of a concept" or opinion of sensation, processing this information seems to take a "lot of effort."[7] And whereas pure monochromes—"Blue in particular" as "coloring void"—are a nexus for force,[8] Deleuze and Guattari suggest that "the most baleful forces" are to be found in the zones of indiscernibility of Bacon's paintings.[9] In effect, the conversation shifts in order of magnitude from conceptual art to figurative painting through the suggestion that the abstract void of either mode of expression is more successfully implemented by Bacon than Klein. This hierarchizing valorization of one mode of expression over another must be acknowledged since it is the engine that generates the momentum of some of the contextually understood concepts that propels them to a mode of expression altogether alien to Deleuze and Guattari.

Let us further explore this notion of the void in relation to abstraction. In fact, it is a figurative void best deciphered through cartographic filters. Deleuze associates cartography with the diagram in painting: "There is a diagrammatical line of desert-distance, just as there is a diagrammatical patch of gray-color, and the two come together in the same action of painting, painting the world in Sahara gray."[10] The diagram is geographical when desert traits make up the painting. This is the desertification of the space of the face in the portrait. What kind of map is part of painting? If the painting were to be completely covered, the desert could turn into marshland: "It must not cover the entire painting, which would be 'sloppy' (we would once again fall into an undifferentiated gray, or a line of the 'marshland' rather than the desert)."[11] This is the geography of the zone of scrambling, the zone of indiscernibility, a space within the diagram. To the desert and the marshland, add the ocean: "The head is split open by an ocean."[12] Of course, this geography relates directly to the type of art Bacon produces: faces smudged, swashed, smeared, but never becoming wholly abstract. Like the map on the wall of a Vermeer interior, Deleuze allows small geographic incursions into the painting but never an entire desert (one can think of Clifford Still's Abstract Expressionist paintings, which could be interpreted as representing vast American deserts), or marshland (Robert Motherwell imagined Pollock's works as teeming waters with an opaque surface),[13] or ocean (William Baziotes created abstract underwater seascapes). The map here is haptic, namely, visual and tactile—like the smudges in Bacon's figures—but not too tactile or "the eye [would have] difficulty following it."[14] Deleuze's assessment of abstraction in the field of art cultivates a careful avoidance of extremes. The map cannot be too geometrical, either, since that would render it too digital, too coded. But it is precisely this idea of code in art that is of interest here. This idea of code is generated by the hierarchizing engine in Deleuze and Guattari's system of art. The nonspontaneous, totally abstract, conventional art that Deleuze cannot get behind. What happens when one zooms in too closely on a digital map? The screen blanks.

Resnais: Superimposed maps

In cinema, like in painting, the map is described in close proximity to the diagram. The diagram is an abstract process between actualized structures, anticipating and creating potentialities. The diagram is dynamic, nonrepresentational, and generative, articulating incorporeal forces. Here, my focus is on the more static, more recognizable map—the map that is traceable and visual. In the context of cinema, however, the map is far from being a static, singular object. Cinema is, after all, a dynamic medium dependent on time. And, as Deleuze suggests, it needs to be seen through the prism of the diagram. This peculiar cartography sheds light on cinema as an instrument of thought. For Deleuze, cartography and thought are linked in particular in the figure of French filmmaker Alain Resnais.

Deleuze writes that thought is at the center of Resnais's films. We can sense already that the cinema of thought will require more than a flat cartography. The map here, consequently, has a specific quality:

> Each map is in this sense a mental continuum, that is, a sheet of past which makes a distribution of functions correspond to a distribution of objects. The cartographical method and coexistence of maps in Resnais may be distinguished from the photographic method in Robbe-Grillet, and his simultaneity of snapshots, even when the two methods result in a common product. In Resnais, the diagram will be a superimposing of maps which define a set of transformations from sheet to sheet, with redistributions of functions and fragmentations of objects: the superimposed ages of Auschwitz. *My American Uncle* will be a grand attempt at diagrammatic mental cartography, where maps are superimposed and transformed, in a single character and from one character to the next.[15]

Of course, the mental cartography of *My American Uncle* (1980) had something to do with the psychological theory that the film meant to illustrate. The "common product" in question that elicits a comparison to *My American Uncle* is, in fact, Resnais's *Last Year in Marienbad* (1961), made in collaboration (or confrontation, according to Deleuze) with Alain Robbe-Grillet. Deleuze suggests that the two individuals came together from different temporal angles: Resnais worked through sheets of the past, while Robbe-Grillet—who, according to Deleuze, did more than write the script—worked from an arresting, perpetual present.[16] Resnais, accordingly, is a cinematographer of topological spaces, whereas Robbe-Grillet is one of quantum spaces[17] who peaks in the present. Resnais's cinema is even interpreted as a complex of information and data.[18] In this reading of Resnais's films, "feedbacks and failed feedbacks" take the place of the time-based flashback witnessed in Robbe-Grillet's cinematographic work.[19] Deleuze sees the cinema of Resnais spatially. But not in a vertical, illusionistic manner: "The readability of the image makes it as independent of the vertical human position as a newspaper can be."[20] Here, he is adapting Leo Steinberg's orientation of the flatbed picture plane as a nexus for information. The cartographic element closes in on the information image. The screen is not read like an illusionistic window one can walk up to and peer through, like a Renaissance painting, but is a map of information providing directional orientation signs: "The vertical of the screen now has only a conventional meaning when it ceases to make us see a world in movement."[21] Deleuze's analysis of Resnais's film yields to a map-image: the space of the screen is seen statically and flatly, not as depth but as surface. The input of information allows Deleuze's analysis to shift from movie screen to electronic screen. This very stance toward electronic images—which will spoil or relaunch cinema[22]—is cartographic. The screen reads as a map.

Karen Beckman explores the cartographic aspect of Resnais's *Last Year at Marienbad*, following Deleuze's references to maps and diagrams used in the film's production. She argues, similarly to Deleuze, for a superimposition of maps. But here the maps are geographical and not temporal. The film, which refers to a Czech town marked by anti-Semitism and implicitly to redrawn cartographic borders right before the Second World War, was shot in Munich in a case of split-location-induced cartographic trauma following the incommensurability of regions. She borrows the notion of incommensurability from Deleuze, as it applies in this cartographic superimposition of contentious regions as a sort of virtual to the action of the film.[23] What is more, Beckman even uncovers a digital unconscious to the film in the prominently featured Nim game—a mathematical strategy game—through its relation to cybernetics in French culture and the mathematical and binary consciousness exhibited in it.[24] The game is played with a pyramidal configuration of four levels of matches: seven at the base, five at the second level, then three, and finally one at the top. The game, played intermittently throughout the film, has increasingly frustrating outcomes for one player who cannot figure out the trick. Finally, it is not without humor that Beckman mentions the map of scenes that appeared twice in *Cahier du Cinema* in 1961. Even though the map was thoroughly footnoted in order to provide a "key" to the elusive meaning of the film, it had to be reprinted when it was discovered that the original was printed upside down.[25] Here, the orientation of the map of a film seems arbitrary and abstract and further supports Deleuze's cartographic reading of a dynamic medium and a variable view of the map itself.

This cartographic assessment of Resnais's films by Deleuze and Beckman shows a number of different maps that appear in a critical view of the film. The map's conceptual elasticity encompasses many dimensions and serves as a potent critical device.

Kafka: Cartography of intensity

The map in Kafka becomes somewhat disembodied. At the start of *Kafka: Towards a Minor Literature*, Deleuze and Guattari mention the multiple entrances to the castle, the innumerable guarded doors of the hotel in *Amerika*, and tunnels dug by a rodent in the bowels of the earth. These are examples of a first type of map. This first type follows the architectural lines traced in the novels and story. Another side of this map observed by Deleuze and Guattari in *Kafka* is the one that contains a degree of the unformed abstraction, always elusive, of *The Castle* and the tricked map of the burrow (one real entrance and the dream of another). The second type of map is made up of multiple points of entry, connected entrances, crossroads, and galleries creating a complex nonhierarchical network resembling a rhizome—its characteristics provide a multifocal lens through which to examine Kafka's corpus itself. From represented maps to maps of a system of representation, *Kafka* introduces a multidimensional cartography that is then elaborated in *A Thousand Plateaus*. One would expect such maps to be somewhat conventional, directional, and orientational. But this map is intensive. As Deleuze and Guattari explain, "The map is a map of intensities, and the moving frontiers are themselves thresholds *(blocks of intensity)*."[26] Intensity, for Deleuze and Guattari, determines the points of connection between various concomitant points of Kafka's corpus. Continuing with their description of *The Castle*, they write,

> The whole first chapter of *The Castle* works in this mode, from threshold to threshold, from low intensities to high ones and vice versa, part of a cartography that is certainly not interior or subjective but that has definitely ceased to be spatial. The low intensity of the bent head, the high intensity of the head that straightens and the sound that takes flight, a passage from one scene to another by thresholds: a language that has become intense makes its contents take flight in place on this new map.[27]

A cartography that has ceased to be spatial is certainly an arresting image. Here the map does not record physical space, but rather unmeasurable points of coincidence. We see this kind of assessment of the cartography in *The Fold: Leibniz and the Baroque*, where maps of intensities are not flat, and where space is not recorded according to visible points but rather is plotted on that which is not traced: "Maps should not be understood only in extension, in relation to a space constituted by trajectories. There are also maps of intensity, of density, that are concerned with what fills space, what subtends the trajectory."[28] Another way of looking at this is to imagine Kafka's system as a spidery weave hovering above ground. Not a Pollock painting, exactly, but rather Eva Hesse's soft-sculpture *Untitled* (1969–70), which looks like someone picked up the painted lines of a Pollock and pulled them away from the canvas, creating a three-dimensional view of something that once was flat. Deleuze is drawing up a map of intensities, whose distances are not spatial or "determined visually" but are determined by relations of intensity.[29] In fact, Kafka's rhizomatic badger tunnel network seems to fit the bill here as an illustration of the type of beyond-the-visual cartography Deleuze has in mind—variability, polyvocal directionality, and a continuously (smoothly) shifting cartography. The badger burrows, feels through tunnels that are invisible to us.

From these three maps (painting, cinema, and literature), we can see the image of the map as unstable, as intruding into a figure, as a superimposable slice of time, and as intensities and thresholds. It is not that we are imagining three different maps, but rather one map-image that is constantly changing. The map-image has cartographic elements that reflect directly on the nature of the image. And this reflection on the image, when we trace it back to its source, is one of the relationship between image and information. But these maps—Bacon's geographical elements, Resnais's spatialization of time, and Kafka's blocks of intensities—are still quite abstract; none are actual images of maps even though they are map-images. In the work of the next artist—and the one I consider in the first chapter—Vermeer, actual maps are represented and yet they point to a dimension beyond representation.

Vermeer's maps

The map-image will be unevenly displayed in the following pages; sometimes the concept will take center stage as the subject of analysis, sometimes it will work behind the scenes. The first chapter explores the source of the map–art–information kernel that propagates through these pages. What is the relationship of art to the map? What is the shift from analog to digital through the map? And how can the map be utilized as a critical device in the analysis of contemporary art? All these questions will be shown to stem from Leo Steinberg's concept of the flatbed picture plane, which the art historian illustrates with, among other things, a map. For Steinberg, there is a temporal division between the art of the Old Masters—representational, illusory, and dependent on the figure—and the art of the information age—concerned with data, self-reflexivity, and a non-illusory orientation. The flatbed picture plane represents a shift from illusionistic art to an informational art. Instead of looking at a canvas as if through a perspectival window, the painting is like the surface of a desk on which is scattered information in the form of pictures and documents (or at least fragments) that must be deciphered instead of contemplated. Vermeer's paintings encapsulate this relationship between the illusory and the informational, between art and maps. In this chapter, we will see how Svetlana Alpers also picks up on Steinberg's flatbed theory and fills out any missing historical contours—via Johannes Kepler—by looking at the baroque conceptualization of the notion of the image: a self-reflexivity that will be a hallmark of more contemporary artists, such as Robert Rauschenberg—about whom, after all, Steinberg was writing when he came up with the notion of flatbed picture plane. Rauschenberg's canvasses must be read. The heterogeneous fragments of reproduced images (photos, magazines, comic books, calendars, printed fabrics) paint spatters in the style of the Abstract Expressionists whose legacy he was supplanting, and the definitely non-illusionistic treatment of the painting's surface have to be deciphered, analyzed, processed—the message, invariably, is that there is no message, just information (in a movement toward postmodernism, and away from Clement Greenberg's modernist formalism against which Steinberg was writing).

Steinberg's discovery of the informational surface of the canvas becomes a plot point in Deleuze's aesthetics. But Deleuze also relies on another seventeenth-century master closer in time to the painter, the philosopher Baruch Spinoza, to advance his survey of the surface of Vermeer's paintings. With Spinoza's help, Deleuze translates the canvas into a screen with plays of light, shadows, and color. Seeing the canvas as screen will transition us to another type of screen: we will see how Vermeer's aesthetics (and the device of the map) enter the cinematographic space.

The map in the painting allegorically points beyond the limits of the frame. In this case, it will point to Ridley Scott's *Blade Runner*, the 1982 science fiction film based on a novel by Philip K. Dick in which Deckard, a film noir–type detective, hunts humanoid robots (replicants) cursed with a short lifespan. Scott appropriates Vermeer's aesthetics (composition, light, palette) in a sequence of the film in which a video device delves into the impossible depths of an evidentiary photograph. But it is also in another of Scott's films—the feminist crime-spree road movie *Thelma and Louise*—that, according to Tom Conley, a baroque use of cartographic details is made manifest. Maps on a motel room wall mediate between screen, viewer, and figure.

For Deleuze and Guattari, the map is an elusive, murky image, often tangentially related to a constellation of geographical notions: strata, territory, smooth and striated space, milieu, and planes. The map-image is also intermedial: a point of passage between different media, in this case, the painting and the film.

Map and code

The second chapter surveys cartography through the concept of the rhizome—which reveals the code element in the duality of mapping and tracing that defines the map's instrumentality for the analysis of visual cultural objects. Yet, recently, the esteem of the rhizome's role in aesthetics seems to have been shifting, or slipping. Andrew Culp, for example, in *Dark Deleuze* agrees that the rhizome would be an apt "snapshot of things to come" except that it lacks the necessary speed.[30] The rhizome advances at a snail's pace, and while some enthusiastically apply it to the Internet as the model of connectivity, Culp warns, "Even something as rhizomatic as the Internet is still governed by a set of decentralized protocols that helps it maintain its consistency—the drawback being that these forms of control are diffuse, not immediately apparent, and difficult to resist."[31] So the rhizome's multiple, dehierarchized nature is what causes it to fall into the hands of a controlling power. Culp suggests that the folding/unfolding model is a superior alternative to the slow and old-fashioned rhizome. I will get to the aesthetics of the folding/unfolding mechanism in the fourth chapter when discussing the virtual/actual dimensions of digital art. The rhizome also possesses connective potential duality. And more specifically, the very protocols, codes, and redundancies that Deleuze and Guattari see as a negative obverse side to the cartographic element of the rhizome will allow us to see code as folded into the rhizome and unfolded unto art.

Another critic of the rhizome model, Timothy Morton, in his dark ecology is also unconvinced of the concept's potential with reference to assemblages.[32] Could we not define even a potato—sometimes bearing fruit, sometimes continuing to grow—as a rhizome? And, Morton asks, what about experimental music claiming to be rhizomatic: Isn't this just a snobbish means of imposing a hierarchical organization where one suspects there is none? And what about when fixing a car using other car's parts: When does it cease to be the original car? Can we ever be at a cusp of two cars? Morton's questions could be answered by reading the cartographic/tracing duality, which aims for a communicable idea based on materiality as the "fantasy" that Morton dismissively accuses the rhizome of being. The rhizome is far more binary than Morton believes it to be, but it is through that trap of binarism that a coded yet cartographic view of the rhizome emerges: A vision no longer of the landscape and models of nature, but rather of representations of space, from the distance of an organizational visual object.

But let us move away from the criticism of the rhizome and focus on the notion of code, which, as I will explain in this chapter, is a necessary facet of the rhizome's aesthetic applications. Aaron Betsky sees code as the fabric of contemporary society. He starts with the "rain" of data that the hero of *The Matrix* (1999) sees as he peels away the curtains of reality. The film is a science fiction commentary about how our current life intertwines with data:

> Computer codes, those vast amalgamations of zeros and ones that control so many of the objects we use and constitute the flow of information that flows all around us, are the most obvious results of the system of combination. They reduce more and more of our reality to a system that is not so much semantic

> as recursive: meaning and action come not out of signification, but out of the abstract relationship of those two digits.[33]

Codes are "generators of banality"—definitely against the jubilation of the rhizomatic map—that produce a sense of alienation between us and our sense of place and its function: "They have reduced the way things work to a level of abstraction we can comprehend but not truly know."[34] But Betsky also sees codes as liberating since—if we think, for example, of DNA as code—they produce more than banality. We are sustained by codes: "In their reductive force, codes clarified, opened up and grounded a new form of knowledge."[35] Art deals with code in different forms: showing it in the quotidian routine; working directly with decoding in order to glitch the code and reveal chaos; or altering physical reality through code, revealing the wonder through artificial means. This is why Betsky has a program to confront code through art:

> By utilizing and revealing both its inherent logic and its artificial nature in a manner that redeems us as witnesses to what is otherwise unpresentable, such an art will make it possible to maintain the sense—if even for a moment and if even in the rarified context of the museum—that code is not only recursive, but discursive as well.[36]

These issues of representability will be treated by Parikka, Galloway, Rancière, and Marks. The definition of code in Deleuze wavers somewhat between the general and the precise; but it always stems from cartography and its obverse—tracing. This is to say that the restricted mission of this book will be to look to art and to the aesthetics of code and its relationship to art.[37]

In the second chapter of this book, I will explore the crucial section of *A Thousand Plateaus* that deals with mapping and tracing. Contrary to a typically jubilant reading of an ever-spreading connectivity, the chapter will examine the redundancies of the rhizome that occur in the obverse of the map. It is through these redundancies that coding comes in, wherein Deleuze and Guattari try to give equal footing to both map and tracing only to stamp out tracing with the far more colorful map. The map folds and unfolds, following the grid pattern indelibly embossed on the paper out of which it arises. A code is necessary for its deployment. The grid holds up the map but is not the point of the map. Even in a Vermeer like *The Art of Painting* (1666–68), the map is part of the gridded, overall composition of the painting.[38] Thus, mapping is not simply the spreading, joyful search for experimentation, but must depend on the tracing. It is a more complicated view of the map than simply saying that it leads to nomadic becomings. The notion of mapping produced by the theory of the rhizome provides the ways of looking at the aesthetics of coding in digital art.

According to Anne Sauvagnargues, Deleuze, like Betsky, looks at different manifestations of code not only as symbolic structures to be decoded, but rather as social productions in the Marxist sense. She explains that Deleuze's notion of the machine as something that cuts and codes is far from the theater machine, which one imagines behind the scenes helping in reproductions of reality. It is rather the machine from a factory, the machine of repetitive, alienating movements. (We will see more of this in my discussion of Harun Farocki's *Eye/Machine* at the end of this book.) Coding, Sauvagnargues continues, "takes a historical thickness as it loses in rationality, becoming a semi-aleatory coding, asignifying rather than following, like Lacan, the ideal of a symbolic structure."[39] The semi-aleatory Markov chain will be the de facto form of the code in Deleuze and Guattari's theories.

And so, code writ-large, through Deleuze's theory and his aesthetic, does not endorse digital aesthetics. Rather, through the two sides of his notion of cartography, we will explore an aesthetic that is not simply optimistic but contends with an uncomfortable relationship between code and culture. To do so, we will analyze Parikka and algorithmic culture; Galloway and the connection between the Internet and mapping; and finally, Marks and the relationship between abstraction and algorithm. Ultimately, Chapter 2 will end with a comparison of the use of code between the analog work of Francine Savard and the digital art of Emmanuel Lagrange Paquet. There is no pre-established relationship between these two artists. I have chosen them to demonstrate the movement from a traditional medium depicting maps and tackling information in Savard's paintings to Lagrange Paquet's software art capturing the cultural impact of cartography.

Mapping intermediality

In the third chapter, I will be mapping intermediality in Cory Arcangel's film-based works. Arcangel is a multidisciplinary artist who plays with media seemingly indiscriminately: video games, cat videos, music videos. One small part of Arcangel's artistic practice is to explore the relationship between the digital and more traditional media such as film and painting. The translation between media is sometimes literally taken, as in his *Translation Exercise* (2006) project, which retranslates an English film into English via a Bangalore call-in center and, in the process, illuminates global networks of capitalist exchange and cultural codes. Arcangel's *Colors* (2006) is a rumination on minimalist painting, codes of the street/society, and information theory. Deleuze is important here because his theories negotiate between digital, painting, and structure. The poetics of the algorithm is present in the paradoxically stark piece *Structural Film* (2007), which shows a blank screen but illustrates the poetics of the algorithm through glitches in the system of translation. Negotiating between the appropriation strategies, the intermedial substitutions, and engagement with a multilayered digital cultural and historical context, a translation theory is necessary, and especially one that relies on Deleuze's geological concept of stratification. Mapping here is implicit: in negotiating dimension, boundaries, and strata, one needs to deploy a stratified spatialized strategy of code culture to access the different layers of representation that reside between expression and theory. Chapter 3 weaves Deleuze's notion of stratification through Paul de Man's deconstruction of Walter Benjamin's translation as nontranslation theory. In turn, these notions of translation lead into theories of information and Deleuze's ideas about the digital and code in art. Finally, we turn to intermediality as the zone of indiscernibility that we already located in the stratification of translation. Even if the notion of mapping is not made explicit, the way this chapter engages with code, media, and milieu—and therefore catches the traits of the map-image—will be instantly obvious.

Virtual art and space

Chapter 4 turns to John F. Simon Jr.'s posthuman aesthetics. Simon's works are algorithmic conceptual art. They show abstract spaces in perpetual mutation that nevertheless obey set parameters. His works also explore philosophical questions. *Swarms* (2002) seems particularly Deleuzean in its theme: Do thoughts obey the swarming motions of animals? The link between animal, thought, and multiplicity is worked out on a double screen, imitating the two hemispheres of the brain, and formalistic color patterns agglomerate and disband. But this work is not Deleuzean only because of the assembling and disassembling animal-like multiplicity; rather, the swarm is a direct line to the virtual in Deleuze: the virtual as the milieu of the agentless posthuman. This is compelling since Deleuze conjures the term "swarm" to explain the virtual mechanism in baroque thought. Simon's work is a portal to figuring out the metaphorical link between the virtual/actual and the algorithm/image dualities. Simon's *aLife* (2003) puts into motion various diagrammatic worlds that show various visualizations of data that are beyond the grasp of the technologically unassisted, or naked, human eye: atoms, molecules, a bird's-eye view of geographical lands, and meteorological formations. The work is a reflection on the nonhuman image: Do environments need to be visually consumed as being meaningful? Are certain images not meant to be visually appreciated? This work opens up reflections on the speed and infinitesimality of information. Finally, Simon's schematically rendered gridiron-patterned New York, *ComplexCity* (2000)—inspired by Piet Mondrian—is a dynamic map. It is an active engagement with abstract painting, code, and the virtual as the grid organizing the work underlies not only urban structures but also the very soul of formalist art.

Picking up on Laura U. Marks's theories about virtuality and algorithmic art, Chapter 4 provides further elaborations on the use of the virtual/actual binary to understand digital art. An exploration of the virtual through Deleuze's use of the term in *The Fold* will be crucial for creating a theory of an agentless operative aesthetic in algorithmic art. I will also look at Gilbert Simondon's theory of the intra-perceptive image in order to discuss posthuman elements of aesthetics and how they can help us define new types of images in the context of a digital environment. Finally, I will analyze the relationship between computer aesthetics stretching back to Mondrian's systematic abstract painting according to Rosalind Krauss's theory of the grid, reflecting back on Simon's digital art and Deleuze's theory of the virtual.

Island/Image apparatus

Chapter 5 maps a series of islands as they are represented in different media (Internet, project for a sculpture, film). Each island of this tripartite archipelago—or aesthetic network based on the idea of the apparatus—illustrates a different manifestation of the digital map-image. First, Janice Kerbel's *Welcome to Bird Island* (2001), a tropical island website, is a reflection on information and the Internet—this map is interactive, informational, and a fictitious document. The second island is Aram Bartholl's *Dust* (2011), a plan for a yet-unrealized sculpture of the virtual space of a video game. Even though the game space being rendered materially is not a geographical island at its source, it nevertheless gains the appearance of an island through its visual characteristics: cut off from a context, from an extended space, it is insular, self-contained, and habitable even though not yet realized. This map-image helps us reflect on screen experience and architectural interface. Finally, the third island is a space capsule digitally rendered within a dystopic science fiction film about space and maps. Ridley Scott's *Prometheus* (2012) plays out space horror conventions by relying on the device of mapping, diagramming, and schematizing interfaces. This holographic map-image exemplifies the notion of cartographic film.

The image of an island, with its creative conditions, is situated in Deleuze's philosophy as a proto-spatial aesthetic theory. Yet, there seems to be an anti-utopian mechanism implanted in the idea of the island, keeping its ideal form at bay. Chapter 5 navigates theories of Deleuze's desert island, Sloterdijk's notion of isolation, Virilio's dromoscopy and bunker theory, Conley's cartographic cinema, and concludes with Agamben's apparatus. The notion of interface is critical, as is virtual space.

Surveilling

The sixth, and last, chapter of the book retraces the steps of the map-image by consolidating the posthuman threads that were discernible in the previous chapters. By focusing on Deleuze's concept of control, this chapter yields a final reflection on the aesthetics of code, the digital, and algorithm by looking at three different allegories of maps. The first one, by Brian Massumi, illustrates the concept of embeddedness of code; the second, following an example by media theorist Günther Anders, illustrates the idea of network control; and finally, Marshall McLuhan's allegory of Roman slaves brings out the realization of obsolescence, or the idea that humans are no longer necessary in the representational universe of digital images. The notion of surveillance captures all three elements and is explored through Hajra Waheed's collages and Harun Farocki's films at the opening and closing of this concluding chapter. These artists zoom in on two new types of images that have emerged out of the digital society: the overhead image (to borrow a phrase from Lisa Parks) and the operative image, whose function does not depend on human eyes scrutinizing and interpreting it.

The following pages, then, will trace an image of the map as it figures in Deleuze's (and Guattari's) corpus and endeavor to showcase its critical versatility in the context of art history and contemporary art.

Notes

1 "In The Postmodern Condition: A Report on Knowledge, Jean-François Lyotard says this text provides 'a layman's version of the de facto impassibility of ever achieving a complete measure of any given state of a system' (55). The desire to achieve complete control over any system, Lyotard says, is thus revealed to be illusory. In the opening paragraph of his essay entitled 'Simulacra and Simulations,' Jean Baudrillard uses the same text by Borges, which he calls 'the finest allegory of simulation' (166), to make a related point about how our century's understanding of knowledge has changed." See Chris Kearns, "Rereading Borges Rereading: A Question of Marginality," *Symplokē* 1, no. 2 (1993): 122n2. Guy Debord, in *Society of the Spectacle* finds that "the spectacle is the map of this new world, a map which exactly covers its territory. The very powers which escaped us *show themselves* to us in all their forces." See Guy Debord, *Society of the Spectacle* (Detroit: Black & Red, 1983), §31. Of course, we will be looking into Deleuze's use of Borges's parable within these pages.

2 Gilles Deleuze, *Francis Bacon: The Logic of Sensation*, trans. Daniel F. Smith (London: Continuum, 2003), 104.

3 Jorge Luis Borges, "On Exactitude in Science," in *The Aleph* (London: Penguin, 2000), 181.

4 Ibid.

5 Gilles Deleuze and Félix Guattari, *What Is Philosophy?* trans. Hugh Tomlinson and Graham Burchell (New York: Columbia University Press, 1994), 181.

6 Ibid., 198.

7 Ibid.

8 Ibid., 181.

9 Ibid., 182.

10 Deleuze, *Francis Bacon*, 158–59.

11 Ibid., 159.

12 Ibid., 100.

13 See Catherine Craft, *An Audience of Artists: Dada, Neo-Dada, and the Emergence of Abstract Expressionism* (Chicago: University of Chicago Press, 2012), 191n306.

14 Deleuze, *Francis Bacon*, 104.

15 Karen Beckman, "Animating the *Cinéfils*: Alain Resnais and the Cinema of Discovery," *Cinema Journal* 54, no. 4 (2015): 1–25.

16 Gilles Deleuze, *Cinema 2: The Time-Image*, trans. Hugh Tomlinson and Robert Galeta (Minneapolis: University of Minnesota Press, 2001), 103, 105.

17 Ibid., 129.

18 Ibid., 266.

19 Ibid.
20 Ibid.
21 Ibid.
22 Ibid., 267.
23 Beckman, "Animating the *Cinéfils*," 19–20.
24 Ibid., 23.
25 Ibid., 17.
26 Gilles Deleuze and Félix Guattari, *Kafka: Toward a Minor Literature*, trans. Dana Polan (Minneapolis: University of Minnesota Press, 2003), 78.
27 Ibid., 78.
28 Gilles Deleuze, *The Fold: Leibniz and the Baroque*, trans. Tom Conley (London: Athlone, 1993), 64.
29 Eugene B. Young, "Haptic," in *The Deleuze and Guattari Dictionary*, ed. Eugene B. Young, Gary Genosko, and Janell Watson (London: Bloomsbury, 2013), 112. He is referring to smooth space.
30 Andrew Culp, *Dark Deleuze* (Minneapolis: University of Minnesota Press, 2016), 38.
31 Ibid.
32 Timothy Morton, *Ecology without Nature: Rethinking Environmental Aesthetics* (Cambridge: Harvard University Press, 2007), 52–53.
33 Olafur Eliasson and Aaron Betsky, "Let It Rain: The Art of the Code," in *Studio Olafur Eliasson: An Encyclopedia* (Köln: Tachen, 2008), 89–90.
34 Ibid., 90.
35 Ibid., 91.
36 Ibid., 92.
37 Niklas Luhmann, *Art as a Social System*, trans. Eva M. Knodt (Stanford: Stanford University Press, 2000), 118. One should not forget Niklas Luhmann's code-based definition of art: "This is why, as we shall elaborate below, a 'code' emerges in art, a continuously maintained binary orientation concerning the 'fit' or 'lack of fit' of forms. This is why every artwork contains 'information' in Gregory Bateson's sense—differences that make a difference … And all of this holds for any kind of art!" Luhmann's theories are an important lens through which to investigate the notion of code and art.
38 Christine Buci-Glucksmann, "Of the Diagram in Art," trans. Josh Wise, *ANY: Architecture New York* 23 (1998): 35: "But by the same token, we find a diagrammatic activity in the work of a painter such as Vermeer, whose grids were made by crossing strings attached to fixed points on a horizon line." See also Phil Steadman, *Vermeer's Camera: Uncovering the Truth behind the Masterpieces* (Oxford: Oxford University Press, 2002), 28: "On the other hand, such 'photographic perspective' cannot in itself be adduced as unequivocal evidence

for use of the camera obscura. It could in theory be the result of Vermeer using quite different aids to achieve geometrically correct perspective. For example, he might have viewed the scene through some kind of squared grid of strings or wires, following the principle of Alberti's 'veil' or some of the perspective devices described by Dürer. He might, indeed, just have followed rigorously the geometrical rules of perspective construction."

39 Anne Sauvagnargues, *Deleuze et l'art* (Paris: PUF, 2005), 128, my translation.

1

Deleuze's Vermeer: Maps, Art, and Information

In this book, I want to train my focus on the aesthetics of Deleuze's amorphous map and how it can be applied to digital art. But how to get from one point to the other—from cartography to digital art? Or, more precisely, is there even a link between the two? To begin, we would say that the map is the emblem of art that deals with information. It seems retrograde to look to Vermeer, a baroque painter, to explore notions of information, digital representation and new media, but Deleuze's approach to Vermeer opens virtual dimensions in selected details of his paintings, such as the wall-mounted maps, the screen-like blank walls, and the use of light. His reading of Vermeer yields a rich constellation of concepts that clarify contemporary aesthetic problems. Most important, through Vermeer, Deleuze discusses the shift from representation to information that occurs in art and leans heavily on Leo Steinberg's notion of the flatbed picture plane to expound on this art historical occurrence in various contexts in his writings. Steinberg immediately elucidates the relationship of the flatbed to the map, and Deleuze, in turn, makes use of this notion to explore the relationship between painting, cartography, and philosophy, especially in his analysis of Spinoza, as we will see. Deleuze's intermediation stemming from Vermeer's maps can be applied to Ridley Scott's *Blade Runner* (1982), where the seventeenth-century painter's imagery appears in the depth of a monitor screen image. Furthermore, Tom Conley's reading of another of Scott's films, *Thelma and Louise* (1991), reveals the baroque imagery of maps that fold films into paintings into a screen.

Flatbed picture plane

Deleuze often circles back to a shift in art that Steinberg pinpointed in his 1972 essay "Other Criteria" (as we have already broached with Resnais). In a decidedly analog medium, the map stands at the crossroads between the window and the screen, between nature and information. Deleuze uses Steinberg's discovery to situate traditional aesthetics in a postmodernist context and, at the same time, charges the concept of the map with a complex aesthetic function. Deleuze writes about Steinberg in *The Fold* (The dyad of the city-information table is opposed to the system of the window-countryside.[1]), *Cinema 2* ("And the screen itself, even if it keeps a vertical position by convention, no longer seems to refer to the human posture, like a window or a painting, but rather constitutes a table of information,"[2] so that "the brain-information, brain-city couple which replaces that of eye-Nature."[3]), and *What Is Philosophy?* (In this case, Deleuze and Guattari bring together Borges's "On Exactitude in Science" and Steinberg's notion of the flatbed picture plane: "The ground covered by its own map, disused spaces without architecture and the 'flatbed' plane."[4]). Although the window/screen duality has been firmly established as belonging to Steinberg,[5] in "What Children Say" Deleuze attributes the duality between movement away from painting as window and its conceptualization as a flatbed of information ("an arrangement … on the surface") not to Steinberg but rather to Svetlana Alpers.

In fact, the specific passage Deleuze quotes from Alpers reads as follows:

> Mapmakers or publishers were referred to as "world describers" and their maps or atlases as the world described. Though the term was never, as far as I know, applied to a painting, there is good reason to do so. The aim of Dutch painters was to capture on a surface a great range of knowledge and information about the world. They too employed words with their images. Like the mappers, they made additive works that could not be taken in from a single viewing point. Theirs was not a window on the Italian model of art but rather, like a map, a surface on which is laid out an assemblage of the world.[6]

Alpers bestows on painters the duties involved in the capture of knowledge and information. She points out that the many discreet sources of visual information have to be added together on a single surface in order to create a work that, even if it gives an impression of illusion, is obviously constructed. And even though Steinberg is not directly cited in this passage, the dualities between illusion and information, window and surface, are clearly there.

It is in another passage of her book that Alpers refers to Steinberg. First, she establishes Johannes Kepler's role in the objective definition of a picture based on his early seventeenth-century research in optics: "It was Kepler who for the first time turned away from the world to a representation of it, to a picture of it on the retina. In structural terms, Kepler not only defines the picture on the retina as a representation but turns away from the actual world to the world 'painted' there."[7] The issue then becomes about two ways of making pictures of the world: "On the one hand the picture considered as an object in the world, framed window to which we bring our eyes, on the other hand the picture taking the place of the eye with the frame and our location thus left undefined."[8] It is this question of the body's location in relation to the frame of the painting that brings Steinberg's flatbed concept into the conversation. The position of the viewer's body in front of the frame is the springboard from which to consider how the surface of the picture is to be understood and why Vermeer is such a compelling artist in the division of painting versus mapping. Steinberg's notion is operative in the window/screen analogy that Deleuze borrowed from Alpers.

The bed and the window are the horizontal and vertical axes determining the orientation of the surface of the painting transposed in Steinberg's "Other Criteria," a response to Clement Greenberg's formalism. The painting, hanging on a wall vertically, like a window opened onto nature, will be given horizontal qualities, like the ones printed on a broadsheet with the help of a flatbed printing press. Or rather, Robert Rauschenberg's *Bed* (1955), a once horizontal object made to accommodate human horizontal activities, is modified (splashed and slathered with thick paint) and hoisted vertically on the wall: "There, in the vertical posture of 'art,' it continues to work in the imagination as the eternal companion of our other resource, our horizontality, the flat bedding in which we do our begetting, conceiving, and dreaming."[9] This horizontal to vertical reorientation is crucial to look at art not as an engine of illusions, but as a repository for information. Steinberg indicates the changeover from a regime of seeing that has been a constant (marked by the vertical orientation of painting starting with the Renaissance and perspective as window onto the world) to a regime of producing and reproducing (as suggested by the three aforementioned activities safely undertaken horizontally). The flatbed picture plane comes from the term "flatbed printing press," which is "a horizontal bed on which a horizontal printing surface rests."[10] Or, as specified by Rosalind Krauss, "a *flatbed* being the horizontal metal tray onto which lead strips of linotype are gathered, before entering the downward pressure of the press."[11] This is a reorientation of how we approach painting, both physically and conceptually.

If we follow Steinberg's timeline, nothing much changed in the span starting with the Renaissance and ending with Abstract Expressionism. Even if the figurative content dissolves into abstraction over the centuries, what stays the same is the way the picture plane in its vertical orientation is parallel to human posture. The vertical human posture grounds the spatial orientation of the picture plane: "A picture that harks back to the natural world evokes sense data which are experienced in the normal erect posture."[12] Consequently, for Steinberg, abstract painters such as Morris Louis—whose veils are suffused in paint dripping downward under the influence of a gravitational force that also naturally affects our bodies—and Jackson Pollock—whose paintings "cannot escape being read as thickets"—are essentially "nature painters."[13]

Deleuze agrees with this reading by blurring the limits of abstraction and figuration in his own assessment of Pollock's abstract paintings, where "chaos is deployed to the maximum. Somewhat like a map that is as large as the country, the diagram merges with the totality of the painting, the entire painting is diagrammatic. Optical geometry disappears in favor of a manual line, exclusively manual."[14] The drip paintings are a manual, physical, part of nature. We are also circling back to Borges: but obviously, here, the map is far from the informative tool of cartography since it is an exact replica of the terrain it is supposed to represent.[15] This particular map does not provide information. It is the abstract machine that has not been concretized into something articulable. Just as Steinberg sees the Abstract Expressionists as nature painters, Deleuze follows the same train of thought by associating them with a pre-modernist, traditional style: "From this point of view, we can see how abstract painting remained figurative, since its line still delimited an outline."[16] Both for Steinberg and Deleuze, Pollock's drip lines are figurative, natural.

For Greenberg, by contrast, the line leads abstraction into the modern era. This notion of the line as tracing an outline is precisely one of the symptoms of the self-reflexive "Kantian" modern painting that Greenberg describes: "Line, which is one of the most abstract elements in painting since it is never found in nature as the definition of contour, returns to oil painting as the third color between two other color areas."[17] Since the line belongs to art and is not found in nature, it self-reflexively signifies the medium to which it belongs. And self-reflexivity within the domain of art is something that Greenberg associates directly with Kant's philosophy. Greenberg's staunch formalism argues for a self-definition of painting based on Kant's critical philosophy, according to which he sees an incremental flattening of the picture plane from Realism to abstraction. For Greenberg, abstraction is the expression of self-definition of painting, progressing from the Old Masters' illusion-producing perspectival figurations: *Arte est artem celare* becomes *Ars est artem demonstrare*.[18] According to Greenberg, pre-modern painters used their mastery to hide the artistry and create illusory space, whereas modernist painters make their art more obvious and show the materiality of their work. The former create illusory space, whereas the latter focus on concrete space within the limits of the canvas.

The reason Greenberg's logic falls apart, Steinberg concludes, is that he does not acknowledge the content of modern art nor the self-reflexive gesture of the Old Masters. This reorientation away from eighteenth-century spatial definition leads Steinberg to the concept of the flatbed picture plane. He describes why this shift needs to happen. If the vertical picture plane is associated with figuration, human posture, and nature in the works of Renaissance and Abstract Expressionist artists alike, then the flatbed picture plane, introduced by Rauschenberg, brings forth a new era of representation with its horizontal orientation. The flatbed picture plane is equated with surfaces like "tabletops, studio floors, charts, bulletin boards—any receptor surface on which objects are scattered, on which data is entered, on which information may be received, printed, impressed—whether coherently or in confusion."[19] A map is another one of these images—it can be tacked onto a wall but it remains a device defined by its horizontal function.[20] Steinberg explains that a radical and new orientation takes place with Rauschenberg: the picture plane does not refer to a visual experience in nature—a visual experience that is part of the realm of the senses—but rather to a conceptual interaction with "operational processes."[21]

The spatial arrangement of the picture plane brings attention to how information and the production of data are the subject matter of the work. Understood self-reflexively: we are looking not at figurative objects, but at nonfigurative images that point to a conceptual process of conveying information about the very painting that contains these visual objects. The image here is *like* an image in the same way that Kant described his schema. Instead of an image, according to Steinberg, we are beholding its psychic address: "What I have in mind is the psychic address of the image, its special mode of imaginative confrontation, and I tend to regard the tilt of the picture plane from vertical to horizontal as expressive of the most radical shift in the subject matter of art, the shift from nature to culture."[22] Deleuze has also described this approach to the essence of the image within the works of the Old Masters, more specifically, Vermeer. In "What Children Say," Deleuze describes the relationship between cartography and art as the relationship between the image and the painting's surface: "A colored map can be present in painting insofar as a painting is less a window on the world, *à l'italienne*, than an arrangement [*agencement*] on a surface."[23] It is no wonder Deleuze looks to Vermeer's works such as *Soldier and a Laughing Girl* (1658)—a close encounter contrasted with a vast expanse in the guise of a map; *Woman in Blue Reading a Letter* (1663–64)—a visual depiction of the conceptual activity of reading; and *The Art of Painting* (1666–73)—an allegory of painting that plays on several levels of reference to vision: painter painting, window/camera, veil, map. As Deleuze says, painting is distilled to the level of spatial arrangements on a surface. In effect, he is buttressing Steinberg's thesis about the self-reflexive Old Masters (they also minded their surface). In fact, at the time of Vermeer, the map itself had not yet been hierarchically and spatially coded. This lack of directionality suggests an underlying conventional surface arrangement in a specific example: James Welu notes that the map behind *Soldier and a Laughing Girl* is hard to recognize because the orientation of the map is atypical. This particular painting displays a map that underscores the idea of surface arrangement because it is not hanging in the typical north–east–south–west position within the frame. It is a bit like Kandinsky who claimed he came up with the idea for abstraction by encountering one of his own expressionist landscapes standing on its side instead of hanging on a wall with its typical illusionistic orientation. The unfamiliar lines arranged on the surface were read as abstract. Welu, in a touchstone essay about the origins of the maps that appear in the background of Vermeer's paintings, notes: "During this period the designing of maps with north at the top was not yet a standardized practice; unlike today's usage, a map could be arranged with north at the left, right, or bottom, according to the preference of the map maker."[24] The map was essentially a nonstandardized, nonhierarchized plane of information. Deleuze, however, goes one step further, and by attributing the process of arrangement to the surface of a Vermeer, where the map is coextensive with the canvas on which it is depicted, he imparts on Old Masters a diagrammatic conceptual flatbed orientation. This orientation would be further explained by the quasi-indexical aspect of the work because of Vermeer's use of camera obscura:

> The maps in Vermeer's painting, such as the one in *The Soldier and a Laughing Girl*, do lose their rectangular quality to some degree—a distortion that may derive from the use of a lens. In *The Soldier and a Laughing Girl*, the case for Vermeer's use of a camera obscura is further supported by his rendering of the map and its relationship to the figures before it.[25]

The movement from canvas to screen has already been engaged.

The surface, the flatbed picture plane, becomes associated with information, thought, and the conceptual. A simple act of reorienting the plane from a vertical to a horizontal position triggers a shift in the way images redefine representation itself. It is no accident that Deleuze's definition of the plane of immanence will be close to Steinberg's flatbed. He will describe it as a table, a plateau, or a planomenon, as a repository for concepts.[26] This is further underscored by the map in Vermeer's paintings: "At the same time the map's rectangular shape contributes to the painting's dominant pattern of straight lines and right angles—the hallmark of so many of Vermeer's works."[27] The flatbed aspect of the map made coextensive with the white wall is yet another aspect of the flatbed picture plane:

> The wall map in *The Soldier and a Laughing Girl*, besides revealing the objectivity of Vermeer's style, also demonstrates how such maps were displayed during the seventeenth century. Here, the colorful and abstract patterns of land and water, the numerous fleets of sailing ships, and the variety of extremely ornate emblems and cartouches form a pleasing complement to the bare white walls of a simple seventeenth-century interior.[28]

This passage explains the number of heterogeneous elements that make up the plane of a map and the map's relationship not only to an empty wall, but more significantly to an empty space on a canvas. One is reminded of the Ocean-Chart in Lewis Carroll's *The Hunting of the Snark* which is left completely blank, a monochromatic slate framed with vague geographical terms. In the domain of conceptual art, Art & Language comes close to Carroll's blank slate with its *Map of Itself* (1967), in which a grid with the inscription "Map of area of dimensions 12″ × 12″ indicating 2,304 ¼ squares" bears witness to a mute blankness. With Vermeer, the bare white wall acts like a blank screen in a movie theater. It captures the light and is a locus for images enacting a network of relationships.

Vermeer and distance

This location of the bare wall and its relationship to light are explored by Deleuze in "Spinoza and Three 'Ethics.'" Even though Deleuze sees Spinoza's system cartographically through an organized fluvial image observed from a great height ("it is like a river that sometimes broadens and sometimes branches into a thousand streams"[29]), it is rather the more austere grid-like flatness of the cartographic image that dominates Deleuze's reading of Spinoza. This text finds Deleuze analyzing Spinoza through three terms that define his philosophy of art as explained in *What Is Philosophy?*: that is, percept, affect, and concept, but with the order somewhat scrambled. In his essay on Spinoza, Deleuze explains how percept is equated with light for itself; affect is manifest in "signs of shadows"; and concept can be seen "as light as color."[30] Deleuze analyzes Spinoza through optics and brings in Vermeer's treatment of shadows through "complementarity and contrast" to make his case about the nature of concepts, percepts, and affects within visual parameters. What is of interest to us here are three separate things pertaining to Vermeer, who is used as a visual manifestation of Spinoza's philosophy: the physicality of the map; the flat swaths of light; and the perspectival matrix of the paintings.

First, a study of Vermeer can demonstrate that a map is much more than its purely physical representation. In Vermeer's depiction of the map, its physicality is made present. Of course, Vermeer's expert rendition of light is responsible for the continuing gravitational pull of the physical map: "Through his skillful handling of light, he gives substance to the map's physical qualities."[31] Light is responsible for something that is beyond the visual: "Here we can almost feel the map's cracked and varnished surface and sense its weight as it pulls down on its two tiny supports."[32] Light renders the map haptic.

Second, light is central to Deleuze's own description of the relationship between Vermeer's art and Spinoza's philosophy:

> Vermeer is said to have replaced chiaroscuro by the complementarity and contrast of colors. It is not that the shadow disappears, but it subsists as an effect that can be isolated from its cause, a separated consequence, an extrinsic sign distinct from colors and their relations. In Vermeer, one sees the shadow detach itself and move forward so as to frame or border the luminous background from which it originates (*The Maidservant Pouring Milk, The Young Lady with a Pearl Necklace, The Love Letter*). This is the way Vermeer set himself in opposition to the tradition of chiaroscuro; and in all these respects Spinoza remains infinitely closer to Vermeer than to Rembrandt.[33]

In *The Maidservant Pouring Milk*, light spills into the room through a gridded window; the bare white wall, seeming luminescent, contours the figure. Behind the main figure of *The Young Lady with a Pearl Necklace*, the wall is a perfectly white surface on which stand out the gold curtains and the coat, the bright hair tie and ornate wood panel. The wall behind the two figures of *The Love Letter* is busier than the ones of the other two paintings, but a right-angled strip is carved out brightly. The white background is a "thing" one sees: framed, it is luminous in its bareness. To those three examples provided by Deleuze, we can add *Letter by the Window*. The light pours in from the outside and organizes space beyond what is framed. The map functions similarly, suggesting a world out of the frame.

Third, another element emerges in Vermeer's particular brand of baroque intricacy that engages with accuracy and precision, manifest through clear perspectival lines and a subcutaneous matrix: the map generates support for the dominant grid-pattern consisting "of straight lines and right angles."[34] The dominant grid, which guides the spectator's vision inward, like Gilbert Simondon's intra-perceptive image that sometimes takes the form of a grid, serves as a beacon generating a centripetal impulse, a movement toward the center. But this centripetal use of the grid, which self-reflexively repeats the framed surface through the grid, also yields the opposite centrifugal movement with a push that propels the image outward toward the world, beyond the bounds of the frame:[35] "Perhaps Vermeer included maps in his interior scenes to suggest a connection between the figures and the outside world, or for the same reason that cartographic material is included in *vanitas* pictures from this period, that is, to symbolize worldliness."[36] The centrifugal trajectory of the grid is revealed most clearly when Buci-Glucksmann gives Vermeer's grids, constructed out of "crossing strings attached to fixed points on a horizon line," the diagrammatic treatment and shows how this pattern resonates virtually, making connections through the history of art and across topological spaces.[37]

In a strange twist, the maps are used as images to illustrate an allegorical figure of *Lady World* or *Frau Welt*, a figure that since the medieval era stands for worldliness: Frau Welt is seductive from the front and covered with sores from the back. In a series of symbolic substitutions, the globe that was originally on Frau Welt's head has been replaced by a map hanging behind a female figure. A touch of macabre pessimism brushes over the maps in Vermeer's bright rooms. The very presence of the map is entangled in geographical aspects of the surface of the painting. Whether it is an orientation of the painting or the way the map folds the distance within an intimate pairing, we are reminded of the ocean splitting the heads of Bacon's figures. The centripetal/centrifugal function of the grid makes the surface issue a distance issue, the way that the face in Bacon collapses the distance of the ocean. Space, as "distance," is more in line with hodology or the study of paths, where connections are examined: "In Vermeer, for example, the most intimate, most immobile becomings … nonetheless refer to the vast distances [*parcours*] displayed on a map."[38] The grid in Vermeer's painting unfolds space into distance.

This relationship of the image to space is also investigated by Jean-Clet Martin in his *Bréviarie de l'éternité*, where he too examines the relationship between Vermeer and Spinoza. He looks closely at the *Girl with the Pearl Earring*, scrutinizing how her gaze creates a roaming space through the floating desire that emanates from her eyes. He writes about this look as something that, over time, becomes unalterable, fixed. This floating gaze "traversed through the camera obscura and was fixed onto the canvas."[39] Martin compares the spatialization of the look and its immovable nature to maps and diagrams. Those elements resemble the face here, giving a legendary dimension to the painting, just as the girl's gaze does in this particular case. Martin is clearly linking the allegorical aspect of the female figure's face and the map. This juxtaposition of face and map echoes Deleuze's own view of the face as "a surface: traits, lines, wrinkles, a long, square, triangular face; the face is a map."[40] The gaze inscribes the image onto the world, just as the map communicates with the rest of the world. The act of getting very close to the canvas itself and seeing the many paths the cracked paint takes to resemble a map seem to produce a zone of indiscernibility. Martin wonders about the nature of this zone in which images appear: the maps and faces trace a trajectory out of the canvas and into eternity, to use Martin's term.

Staring at the map in Vermeer pushes us to other indeterminate dimensions. Maybe this is why we encounter Vermeer again, hundreds of years later, in a future yet to come. In Ridley Scott's *Blade Runner* (1982), a sequence of an uncanny Vermeer-like quality unfolds in which a photograph is explored by Deckard (Harrison Ford), the blade runner detective assigned to destroy rogue humanoid robots (replicants) that have escaped off-world colonies and returned to earth seeking to extend their artificial lives. To analyze a photograph of a hotel room, Deckard uses an electronic device, the Esper machine, tracing a virtual pathway starting on the surface of the photo and continuing a trajectory behind the walls and through reflections in mirrors to come to rest on the coded number inscribed onto a snake's scale.[41] Vermeer's aesthetic is framed and conjugated through the screen: a photograph frames it, we traverse its virtual space through the television monitor's screen, and, of course, ultimately it is all projected onto a screen in a movie theater. It does seem that the space covered in the Vermeer that Martin describes functions like the expansive space of the Vermeer-like photograph of *Blade Runner*.

The Vermeer photograph is also associated in *Blade Runne*r with coding and digital control. The photo yields a clue: a snake's scale with the serial number XB71. This is the kind of coding and control Deleuze has in mind:[42] every scale is tagged, every code leads to a dealer and a customer. Deckard creates a path in the city that starts in the photograph and snakes its way, as it were, through the city, ending in a final burst. Deleuze, after all, declared the snake the emblem of the society of control.[43] The replicant that Deckard guns down was already marked for death, a lifespan coded into her DNA. Of course, when the replicant leader, Roy Batty (Rutger Hauer), dies at the end of the film, the outer worldliness is made clear: "I've seen things you people wouldn't believe." The distance is outside of earthly human experience. The light of the morning brightens the death scene.

The light also comes into play in coloring the celluloid Vermeer in the hotel image seen through the Esper machine described above. Stephen Mulhall, in his analysis of the Esper machine sequence, describes the impact of light on the room in the photograph. The light and shadows play a role in veiling and revealing a virtual space within the logic of the film by closely linking it to a painting and its use of color. In this case, the light, as described by Deleuze and Martin, is featured prominently in Mulhall's description of the film:

> The scene is illuminated by sunlight shining through a pair of windows in the left-handed wall of the first room. It is cast at a shallow upward angle (so the sun must be low, and the room most likely high in the hotel). It reaches across the whole space, and gives a quietly magical richness and depth to the various everyday things it touches, conveying the volume of space that each occupies, capturing the substantial transparencies of glass and reflection, and distinguishing clearly between subtly varying shades of cream, very light green and brown. But his light also casts strong shadows, so we do not immediately realize that, at the table on the left near the window, we can see the shoulder and arm of a man whose head is turned away from us and is anyway (along with the rest of his body) in starkly contrastive darkness—someone as still and substantial as any of the objects surrounding him: just one more solidly material, quotidian thing on which the setting sun casts its indifferently transfigurative light.[44]

This description of the scene as the interplay of light and shadows, space and colors, with the light coming through off-scene windows and casting shadows and darkness, is a summary of the three concept-types Deleuze finds in Spinoza—light for itself, shadow signs, light as color—here garnered to establish a slice of the fugitive life of replicants holing up in a hotel while at the same time articulating the intensive point of view of a machinic eye.

The hotel room as a setting for affective maps is picked up again by Scott in another film that displays a scene of momentary respite for fugitives in an infinite trajectory. In *Cartographic Cinema*, Tom Conley finds a connection between mapping, Vermeer, philosophy, and Ridley Scott. Conley, who has written extensively on Deleuze's cartographic methods, explains how Scott, in *Thelma and Louise* (1991), a film far from the science fiction genre, creates a Vermeer-like interior in a roadside motel with a Dutch mirror and maps on the wall. Here, too, as in Vermeer and in the photo in *Blade Runner*, the sunlight touches the figures in the most painterly way in a setting that inspires reflection on "art, cartography, and cinema."[45] Conley also comments on Alpers's "pathbreaking" monograph on Dutch painting, quoting the same passage that was of interest to Deleuze, and reflects on Steinberg's connection between painting and knowledge. Conley's description of the room in *Thelma and Louise* has something of a baroque complexity. He walks us through the many details and virtual dimensions, similar to the Esper device from *Blade Runner* (Both *Thelma and Louise* and *Blade Runner* display tawdry temporary accommodations as hardened Dutch interiors; even the Van Eyck mirror of *Blade Runner* could be compared to the "curvature of the television screen" of *Thelma and Louise.*[46]). Perhaps Thelma's position in front of a map marks her allegorically as a Frau Welt. Conley ends his analysis with an outward/inward look, like that described by Welu and Martin in their respective analyses of maps in Vermeer: "The women in Vermeer … seem to live in full cognizance of the greater space and light infusing them with sensuous grace. They turn the places where they are into matrix-like space whose defining surfaces promise depiction of greater worlds to come. Their rooms are pregnant with potential of growth and inner travel."[47] This summarizes the theme of space within the Vermeers: a space of intensity that seems to point outward through a centripetal/centrifugal mechanism based on the degrees of self-reflexivity emanating from a core of superimposed maps mediating between picture frame, viewer, and figure. This started with Deleuze's assessment of Vermeer's work as an arrangement on the surface, the flattening out of a three-dimensional illusory space triggered by the presence of the map.

Perhaps we can end the path from painting to screen, from art to information, with Deleuze's assessment of the modern baroque. The thread is from one facet of the classical idea of baroque (Vermeer) to a comparison with Abstract Expressionism and the reversal of the surface of the painting (Steinberg) to finally a neo-Dadaist usage of the painting's surface as a matrix of information (Rauschenberg). In *The Fold*, Deleuze explains the relationship between the outside and the window and how this refers to painting; this is the relationship between light and surface that we have been reading in Vermeer: "A painting always has a model on its outside; it always is a window."[48] But Deleuze makes the connection to a modern image, a moving image (validating, on some level, the comparison between Vermeer's paintings and Scott's films): "If a modern reader thinks of a film projected in darkness, the film has nonetheless been projected."[49] A third step is to move from painting to film screen to computer screen: "Then what about invoking numerical images issuing from a calculus without a model?"[50] In three steps, Deleuze moves from the baroque to the algorithmic image. And, of course, the surface of the painting is occupied by a line, one that threads the space of the canvas onto an infinite outside: "Or, more simply, the line with infinite inflection that holds for a surface, like the lines of Pollock's or Rauschenberg's painting?"[51] Or the infinite paths of Vermeer's maps. Following Steinberg, this passage of *The Fold* reiterates the relationship between art and information, between painting and the code image: "More exactly, in Rauschenberg's work we could say that the surface stops being a window on the world and now becomes an opaque grid of information on which the ciphered line is written."[52] The movement from analog painting to digital art is achieved by following the groundwork laid out by Steinberg: "The painting-window is replaced by tabulation, the grid on which lines, numbers, and changing characters are inscribed (the objectile)."[53] This is why even though the map-image is initially rooted in the information flatbed described by Steinberg, its multidimensional dynamism is captured through an intensive mesh that brings the map closer to a fluctuating cartographic image. The map depends on its surrounding milieu.

Conclusion

Overall, the map-image covers several ways in which geographical or spatial images are used throughout Deleuze and Guattari's corpus in relation to aesthetics. Specifically, the relationship charted is one between information and art. Since more and more art depends on information—as a method, as content, as form—I propose here the map-image as the guiding concept. This is just one example of how the map becomes a path from the canvas to the screen. Vermeer is the ideal guide for this trajectory, starting with a self-reflexive surface of an analogical medium and ending with a cinematographic screen with implied digital dimensions. In a previous monograph, I explored the notion of the philosophically complex diagram, which utilizes visual strategies to enact a non-visual image of thought's processes. The diagram's function was one of transition from one system to another. And it seems like it would be an apt tool to examine which traits migrate from painting to cinema, from one medium to another. So why revert to the map? Because the map as a concept seems to be inherently more representational. That is why the concept of the informational image originates with Steinberg's art historical assessment of the flatbed picture plane and is incorporated at various times by Deleuze into his philosophy to explicate abstract notions of digital aesthetics. In fact, with Steinberg's concept in Deleuze's philosophy, we can dislodge art from the domain of representation, reorienting the spatial layout of the picture plane toward processes of code operative in the image.

Notes

1 Deleuze, *The Fold*, 30.
2 Deleuze, *Cinema 2*, 265.
3 Ibid., 267.
4 Deleuze and Guattari, *What Is Philosophy?*, 168.
5 Gilles Deleuze, "What Children Say," in *Essays Critical and Clinical*, trans. Daniel W. Smith and Michael A. Greco (Minneapolis: University of Minnesota Press, 1997), 191n13; Svetlana Alpers, *The Art of Describing: Dutch Art in the Seventeenth Century* (Chicago: University of Chicago Press, 1983), 122.
6 Alpers, *Art of Describing*, 122.
7 Ibid., 37.
8 Ibid., 42.
9 Leo Steinberg, *Other Criteria: Confrontation with Twentieth-Century Art* (New York: Oxford University Press, 1972), 90.
10 Ibid., 82.
11 Rosalind Krauss, "The Slung Leg Hypothesis," *October* 136 (2011): 221.
12 Steinberg, *Other Criteria*, 84.
13 Ibid.
14 Deleuze, *Francis Bacon*, 104.
15 Borges, "On Exactitude in Science," 118.
16 Deleuze, *Francis Bacon*, 105.
17 Clement Greenberg, "Towards a Newer Laocoon," in *Art in Theory: 1900–1990: An Anthology of Changing Ideas*, ed. Charles Harrison and Paul Wood (Oxford: Blackwell, 1996), 554–60, 558.
18 Ibid.

19 Steinberg, *Other Criteria*, 84.
20 Ibid.
21 Ibid.
22 Ibid.
23 Deleuze, "What Children Say," 66.
24 James A. Welu, "Vermeer: His Cartographic Sources," *The Art Bulletin* 57, no. 4 (1975): 530.
25 Ibid., 531. Although as Alpers states, the attempts at proving such a use of the mechanism are slippery, see Alpers, *Art of Describing*, 29–30.
26 Deleuze and Guattari, *What Is Philosophy?*, 38.
27 Welu, "Vermeer," 533.
28 Ibid., 532.
29 Deleuze, *Essays Critical and Clinical*, 138.
30 Ibid., 148.
31 Welu, "Vermeer," 536.
32 Ibid.
33 Deleuze, *Essays Critical and Clinical*, 143.
34 Welu, "Vermeer," 533.
35 Rosalind Krauss, "Grids," *October* 9 (1979): 61–63.
36 Welu, "Vermeer," 541; "As an image of an image, the flatbed matrix is no longer a world-space." John Cunnally, "Antiquarianism and the Origins of the Flatbed Matrix," *Notes in the History of Art* 31/32, no. 4/1 (2012): 11.
37 Buci-Glucksmann, "Of the Diagram in Art," 35.
38 Deleuze, *Essays Critical and Clinical*, 66.
39 Jean-Clet Martin, *Bréviaire de l'éternité: Vermeer et Spinoza* (Paris: Éditions Léo Scheer, 2011), 49–50. "*La Jeune Fille à la perle*, déposée dans sa brillance si singulière, montre la clef d'un visage, d'un désir flottant, traversant la chambre obscure et fixé sur la toile pour toujours, aussi peu modifiable d'ailleurs que les cartes, les diagrammes et mappemondes que la peintre suspend derrière ses modèles come pour produire une légende, les inscrire dans le tout du monde (nommé Dieu ou la nature). Dans quelle zone flottent ces images qui naissent dans le temps mais semblent devenir éternelles ?"
40 Tom Conley, "Faciality," in *The Deleuze Dictionary*, ed. Adrian Parr (New York: Columbia University Press, 2005), 97, quoting Gilles Deleuze, *Dialogues with Claire Parnet*, trans. Hugh Tomlinson and Barbara Habberjam (Minneapolis: University of Minnesota Press, 1987), 170.
41 Paul M. Sammon, *Future Noir: The Making of Blade Runner* (New York: Harper Collins, 1996), 258: "The whole setup was lit by Ridley to look like the paintings of Vermeer and Edward Hopper."

42 William M. Kolb, "Blade Runner Film Notes," in *Retrofitting Blade Runner: Issues in Ridley Scott's Balde Runner and Philip K. Dick's Do Androids Dream of Electric Sheep?*, ed. Judith B. Kerman (Bowling Green: Bowling Green State University Popular Press, 1991), 162.
43 Gilles Deleuze, "Postscript on Control Societies," in *Negotiations, 1972–1990*, trans. Martin Joughin (New York: Columbia University Press, 1995), 180, 182.
44 Stephen Mulhall, "Zhora through the Looking-Glass: Notes on an Esper Analysis of Leon's Photograph," in *Blade Runner*, ed. Amy Coplan and David Davies (London: Routledge, 2015), 103: "Can a Hollywood science fiction movie really claim to take the measure, or, and so to measure itself against, the kind of artistic ambition exemplified by Vermeer? At the very least, the sheer presence and the unfolding purpose of this photographic mimesis of ordinary life suggests an interest on Scott's part in reflecting on the relative powers and constraints of painting and photography, and thereby on the material basis of his own cinematographic medium."
45 Tom Conley, *Cartographic Cinema* (Minneapolis: University of Minnesota Press, 2007), 169.
46 Ibid.
47 Ibid., 170–71.
48 Deleuze, *The Fold*, 27.
49 Ibid.
50 Ibid.
51 Ibid.
52 Ibid.
53 Ibid.

2

Map and Code in *A Thousand Plateaus*: Savard, Lagrange Paquet, and Data Art

The map-image seeks to bend, twist, and push Gilles Deleuze and Félix Guattari toward an aesthetic based on technology and information. One could argue that Deleuze and Guattari's taste in art veers toward the canonical. In the "Percept, Affect and Concept" chapter of *What Is Philosophy?*, for example, Deleuze and Guattari enumerate mostly Western grand masters of classical and modernist traditions, including Cézanne, Rameau, and Dickinson, among dozens of others. At the end of their assessment of affect and percept in painting, music, and literature, they seem unable to land firmly on the question of abstract and conceptual art. It is odd that philosophers who claim that the problem with the abstract machine is not that it is too abstract but that it is not abstract enough would not fully integrate these types of art into their canon.[1] How would they feel about abstract, digital, conceptual, network, and algorithmic art? The issue arising is how to apply Deleuze and Guattari's conceptual apparatus to artistic objects they did not, or were unable to, fit into their repertoire of examples.

For this purpose, one concept in their arsenal is particularly primed for contemporary art, both as an image and through its function. This concept/image is that of the map. A concept that, following Deleuze and Guattari's definition, has irregular contours; it articulates and cuts and cross-cuts. The irregularity and dynamism that imbue the definition of the concept through the cinematic terms of image editing are helpful here, signaling that the map as concept is not defined as regular and consistent but also that it unfolds in time like a work of cinema. Furthermore, borrowing parameters from Deleuze and Guattari's definition of a concept and applying them to the map, the map-as-concept has totalizing tendencies: even if we have an encompassing idea of its heterogeneous components, the map remains fragmentary and shifting as a whole depending on which facet we approach it through.[2] Like an image that is not quite an image, the map's contours can be seen without it being a representation of a particular location. Our image of the map is creative and not static. The map, then, comes up multiple times in Deleuze and Guattari's writings. In different contexts, it has different functions; but it nevertheless gravitates toward the aesthetic object—the traits that make up its cartographic functions tend to regroup around art. And while the philosophers would not have used it to analyze information-based art, the informational turn away from representational aesthetic is precisely the source of their own mobilization of the term "map."

The map as critical method is readymade for the art produced in the information age. This aesthetic device and its relationship to information come to Deleuze from the writings of art historian Leo Steinberg, who described the map as conceptually parallel to a table of information evoking "operational processes."[3] Deleuze refers to Steinberg in *Cinema 2*,[4] *The Fold*,[5] *What Is Philosophy?*,[6] and *Essays Critical and Clinical*.[7] The map replaces the illusionistic space of a perspectival landscape and provides a data-based treatment of space. But the map as critical apparatus is nimble enough for Deleuze to apply it to several artistic media. A cartographic device can be smoothed over the surface of a painting (deserts in the painting of Bacon), folded into film (cinematic superimposed maps of Resnais), and placed over a text (Kafka's burrow network).[8] Even sculpture gets the cartographic treatment (in "What Children Say," Deleuze explains how psychoanalytic mapping can be applied to an environmental art intervention[9]) as does music ("Of the Refrain" is about creating territories through songs—children's songs, bird songs). But how does one use a map in relation to information? The informational map must engage with the virtual and actual, deal with coding, and use the screen as its interface.

In what follows, I will set out to determine the place of the map in Deleuze's aesthetics and see how elements of this cartographic concept/image can be applied to contemporary digital art (writ large, because even traditional media like painting can be created under some sort of digital influence). First, I will examine Deleuze and Guattari's description of cartography and decalcomania in *A Thousand Plateaus* in relation to their concept of the rhizome. This is the focal point of the cartographic concept, and the means through which it can become a method for the visual, cultural objects insofar as it challenges notions of representation. I will isolate elements in the double principle of cartography and decalcomania with digital resonances—such as redundancy/code—as a way of getting closer to a "digital" Deleuze. Second, I will explore how the code is explained in *Anti-Oedipus* by the Markov chain, in turn cloaked as an aesthetic critical device at the root of Deleuze and Guattari's own multidisciplinary methodology. Third, I will look at the relationship between code, algorithm, mapping, and art in the writings of Jussi Parikka, Alexander R. Galloway, and Laura U. Marks. And finally, I will examine two artists who use cartographic strategies in their artistic practices—Francine Savard and Emmanuel Lagrange Paquet—comparing how they illustrate the shift between code in painting to code on screen.

Cartography and decalcomania as visual representations

The map described by Deleuze and Guattari in *A Thousand Plateaus* is a concept consisting of two sides: on one side, cartography, and on the other, decalcomania. Let us step back: the rhizome consists of many different principles. Two of these, five and six, are cartography and decalcomania. So, in effect, it is the rhizome that operates according to these two principles (among other principles, such as connection, heterogeneity, multiplicity, and asignifying rupture). But in their description of these two principles, it becomes apparent that it is mapping that they hold dear. The map tends to have rhizomatic characteristics, whereas decalcomania does not. Decalcomania, or tracing, is ostensibly the obverse of the map (and by extension all that is rhizomatic). The map, which can be hung on a wall like a painting, and its opposite, decalcomania, or "the art of transferring pictures or design from specially prepared paper"[10] (or, more simply, copying by tracing), are concepts of visual representation. Mapping is always abutting the unknown, the virtual; whereas tracing, a device of repetition and copying, is always necessary to transfer the unknown captured by mapping into something recognizable. What is of interest here in Deleuze and Guattari's definition of mapping is how it relates to code and how mapping generates an aesthetics of code.

The principles of cartography and decalcomania can be understood as representations. They are principles of the rhizome, Deleuze and Guattari's nonhierarchical, multiple, generative star concept. Two manifestations of tracings are genetic axis and deep structure: structuring and generative models in psychology and grammar in which a baseline accounts for variations. The operative model behind tracing is that of reproducibility and repetition. The object of such a model is crystallized as representation and code. It negotiates between fluid relations and describable status quo—it explores something expected. The wasp and the orchid, Deleuze and Guattari's emblem par excellence of rhizomatic heterogeneous connectivity, are the pivot point between tracing and mapping. At first, Deleuze and Guattari advance the possibility that the orchid could imitate the wasp.[11] But imitation quickly cedes the floor to code: "Not imitation at all but a capture of code, surplus value of code, and increase in valence, a veritable becoming."[12] Something new is created by this assemblage; its valence is neither fully wasp nor fully orchid, but a new mapping. The orchid does not reproduce the wasp; rather, a map is formed at their pairing.[13] The map is the creative principle open to experimentations, to new constructions; it does not reproduce objects, but instead engages with the relations between them. And even though decalcomania and tracing have roots in artistic reproduction design, it is the map that is aligned with art in spirit and in form and becomes the site of the interface of rhizomatic connective aesthetics:

> The map is open and connectable in all of its dimensions; it is detachable, reversible, susceptible to constant modification. It can be torn, reversed, adapted to any kind of mounting, reworked by an individual, group or social formation. It can be drawn on a wall, conceived of as a work of art, constructed as a political action or as a meditation.[14]

Here the intermedial interface of the map is apparent. The clearest relation between the map and the rhizome can be seen in the abundant entries the map possesses, much like the burrow of the "animal rhizome"—such as the muskrat, the example provided in *A Thousand Plateaus*.[15]

But "multiple entrances" are already mentioned at the outset of *Kafka*: the multiple yet unlocalizable entrances of the castle and the multiple but guarded entrances of the hotel.[16] The *Kafka* burrow, unlike the *Thousand Plateaus* burrow, has one entryway (and it is used for surveillance), but the very description of the burrow is a trick—an unrepresentable map: "Even when the animal is unique, its burrow isn't, the burrow is a multiplicity and an assemblage."[17] The map, on one hand, is always site of multiplicities, even when there is only one entrance; the tracing, on the other hand, is redundant: "A map has multiple entryways, as opposed to the tracing, which always comes back 'to the same.'"[18] But there is more to the tracing and the map than their ostensible opposition. And again, in *Kafka*, Deleuze and Guattari elucidate the relationship between the tracing (photo) and the map: "One might say that in projecting the photo of the father onto the map of the world, Kafka unblocks the impasse that is specific to the photo and invents a way out of this impasse, putting it into connection with a whole underground network."[19] The tracing and the map create a network. But this is also how Deleuze configured the duality of code and the analog in painting through the work of Bacon, where painting and photography merge: "Bacon explains that when he does a portrait, he often looks at photographs that have nothing to do with the model—for example, a photograph of a rhinoceros for the texture of the skin."[20] And the photograph is associated with code, and painting with analogy.[21] The map-image logic functions in art between code/analogy, painting/photography, figurative/abstract fault lines.

Deleuze and Guattari catch themselves playing up the dualities that the rhizome was supposed to resist. And they do not want maps to be the good side and the tracing the bad side. The map is traceable after all: "Does not a map contain phenomena of redundancy that are already like tracings of its own?"[22] The tracing does not reproduce, repeat, or copy a map; instead, it selects, isolates, restricts a part of the map. A still snapshot of a map in flux. "The tracing has already translated the map into an image."[23] And when Deleuze and Guattari seem to have found something nice, or at least constructive, to say about the tracing, they fall back to casting it as the flipside of the map: the tracing thinks that it is producing something new by selecting elements of the map, reproduces itself, and a feedback loop that is its essence begins to show: "That is why the tracing is so dangerous. It injects redundancies and propagates them. What the tracing reproduces of the map or rhizome are only the impasses, blockages, incipient taproots, or points of structuration."[24] The tracing is responsible for making the map a representational, readable, communicable object. But Deleuze and Guattari's seemingly negative account of the role of the tracing makes a point to reverse the order: Do not take photos of the map, but put the tracing back onto the rhizome: "The coordinates are determined not by theoretical analyses implying universals but by a pragmatics composing multiplicities or aggregates of intensities."[25]

The centralized/decentralized aspects of trees/rhizome images lead Deleuze and Guattari to information and computer science that "grant all power to a memory or central organ."[26] There, the issue is the primacy given to hierarchical structures that link arborescent models to topologies working within predetermined channels. Topology would be replaced by a "transduction of intensive states."[27] If topological stretching or twisting of a particular configuration leaves these unaltered despite their strategic deformation, which results in a homeomorphism that on some level is a form of redundancy, then transduction is a transmission of codes through heterogeneous milieus that alter the function of each according to the traits of the initial milieu. The rhizomatic model, then, is one that is acentered rather than central, a "nonhierarchical, nonsignifying system without a General and without an organizing memory or central automaton, defined solely by a circulation of states."[28] This is not an ondulating homostasis of the same, but an altering force spreading through a network. Deleuze and Guattari refer to the theories of Pierre Rosenstiehl and Jean Petitot, two French mathematicians who advanced a cybernetic chart theory, which, of course, can be read from an aesthetic philosophy point of view as a mathematical mapping theory. But tracing must isolate elements of the map to permit communication, even while proceeding through anexactitude.[29] And so the tension between the map and the tracing, cartography and decalcomania, could be that of representing the unrepresentable. The map actualizes the virtual by recognizable means: "Unlike the graphic arts, drawing, or photography, unlike tracings, the rhizome pertains to a map that must be produced, constructed, a map that is always detachable, connectable, reversible, modifiable, and has multiple entryways and exits and its own lines of flight."[30] A snapshot is required even if it does not fully do justice to the subject matter.

The theme that emerges from Deleuze and Guattari's reading of Rosenstiehl and Petitot's text on acentral systems as networks is that images that side with restrictive hierarchization are seen as arborescent structures and fluvial directionalities that function like topological tracings.[31] These images are there to illustrate the flow of information in societal multiplicities. A weather image, that of the cloud, stands in contrast to the geographical one. In a mosquito scourge, each individual of the cloud formation is constantly correcting its aleatory position in relation to the discreet and immediate placement of the other individuals, essentially operating the vast network through a transductive mechanism. It is not a top-down organization, but a catastrophic stability.[32] We can see the tracing and the mapping exemplified in forms slightly different from those of Deleuze and Guattari. But, and this needs to be underscored, the contrasting images offered by Rosenstiehl and Petitot come from a digitization of information as it is applied to the individual. The problems worked out by the mathematicians need to be represented by numbers, or digits, as they explain, and the shift between states of affairs also needs to be digitized.[33] This marriage between digitization and spatial models is articulated through the relationship between milieus via transduction.

Let us see exactly how the idea of transduction operates spatially. In the chapter entitled "1837: Of the Refrain," Deleuze and Guattari further their definition of code by bringing it together with the notion of mapping and milieu. The chapter opens with three instances of the refrain, but they might as well be three distinctive mapping actions: the first is the child who sings a song to himself to organize the surrounding chaos, to orient himself, and with the song structures a sheltering space; the second map is a home where the space is organized against the outside forces; the third is a portable orientational device open to the outside, to the future, and to the World. Each milieu within chaos is organized and structured according to code. One of the definitions of code is based on the communication between milieus and on differentiating between the decoding and transcoding features of territories. Thus, Deleuze and Guattari explain, the relationship between milieus is based on code. A code, in this case, is part of refrain, repetition, redundancy: "Every milieu is coded, a code being defined by periodic repetition; but each code is in a perpetual state of transcoding or transduction."[34] The code has a function—the function of establishing a milieu but also of communicating with other milieus: "Transcoding or transduction is the manner in which one milieu serves as the basis for another, or conversely is established atop another milieu, dissipates in it or is constituted in it."[35] A map of networks between milieus can be imagined. The term "transduction," indicative of Simondon's influence on Deleuze, designates the transmission of information from one medium to the next. In this case, the information is the code, or the patterns belonging to each milieu, which are then transmitted from one milieu to the next: "The notion of the milieu is not unitary: not only does the living thing continually pass from one milieu to another, but the milieus pass into one another, they are essentially communicating."[36] This notion of coding and transcoding/transduction firmly establishes an information-based image of cartography in Deleuze and Guattari's spatial theory.[37] Which brings us back to the wasp/orchid duality as the crux of the code/map distinction. One of the manifestations of the relationship between milieus is the reciprocal relationship of wasp and orchid (among others, such as spider and fly, snapdragon and bumblebee, and even water and leaf).[38] And, as we have seen, the meeting point between species generates a map.

What is interesting is that coding, which is essentially redundancies, features prominently in Deleuze's definition of abstract art, subordinating, as Eugene B. Young explains, tactile and nonvisual elements to a visual field. Furthermore, a code is a part of the digital that itself is subordinating manual and haptic, or analog, elements in art.[39] The code is also called a redundancy in the specificity of the strata, hand in hand with territorialization.[40] The word redundancy comes from information theory: "In information theory, redundancy refers to the degree of efficiency of message transmission. Human language includes a great deal of redundancy, as compared, for example, to the low redundancy of computer algorithms."[41] Within the spectrum of code as an element of the rhizome/cartography definition, we need to reconsider the modulation of code: "through operations of 'coding', 'decoding' and 'overcoding.'"[42] It becomes apparent that coding is part and parcel of a theory of translation and modulation leading to the haptic, the aesthetic notion of visual prehension Deleuze explains in his book on Bacon, which, in turn, points to the zone of indiscernibility integral in a Deleuzean digital aesthetics.[43] Art is the key to visualizing how milieus combine through code.

Painting and code

Combination is part and parcel of Deleuze's concept of code as it relates to the digital in painting. In fact, we have to take a step back to appreciate the peculiar way Deleuze uses code and the digital in relation to painting.

Deleuze writes about abstract painting as a symbolic digital code. For Deleuze, abstract painting "proceeds by code and program, implying operations of homogenization and binarization that are constitutive of a digital code."[44] This is a problem since painting, in general, is the "analogical art par excellence."[45] "But," Deleuze concedes, "the abstractionists often happen to be great painters."[46] Regardless of the potentially patronizing tone of this sentence, Deleuze implies that it is the greatness of abstractionists that prevents them from mindlessly applying an external code and spurs them to seek internal solutions to the problem of painting. Deleuze, despite his suspicious view of abstract painting, concedes as much. We must say that the digital Deleuze has in mind is not that of the coder but that of the artist. It is an aesthetic code: but one that nevertheless functions restrictively as an algorithmic operation. From the milieu of software development to that of art and visual abstraction, we find a transduction of function in the diagrammatic sense of the term.

The digital Deleuze has in mind is meant in the sense of "a finger that counts."[47] He brings up an example of Vassily Kandinsky's "coding" of the abstract picture plane: "Thus, according to Kandinsky, vertical-white-activity, horizontal-black-inertia, and so on. From this is derived a conception of binary choice that is opposed to random choice."[48] Choice and randomness will be important in relation to the wider notion of coding Deleuze engages with in his philosophical system. Here, what he has in mind is a conventional language created by a contrived and systematic, anti-chaotic abstraction: Auguste Herbin, Kandinsky. Herbin, for example, whose bright-colored abstraction seems ordered with clear geometric shapes, devised an alphabet, or a code, that equated letters of the alphabet with their respective colors, shapes, and musical notes. Kandinsky, whose paintings or compositions analogize between the abstraction of music and painting, wrote several treatises systematizing abstraction. Deleuze does not see code in the best light: "The hand is reduced to a finger that presses on an internal optical keyboard."[49] The code, according to Deleuze, has a direct link to the cerebral: "The code is inevitably cerebral and lacks sensation, the essential reality of the fall, that is, the direct action upon the nervous system."[50] The notion of code and painting come together through an aesthetics of geometry, something measured and controlled. Geometrical uses are digital, or conventional:

> We called one of these uses "digital," not in direct reference to the hand, but in reference to the basic units of a code. Once again, these basic units or elementary visual forms are indeed aesthetic and not mathematic, inasmuch as they have completely internalized the manual movement that produces them. They still form a code of painting, however, and turn painting into a code.[51]

The relationship between code and gesture is paradoxical. Deleuze will show how dualities such as analogical versus digital, right side versus left side of the brain, relational and expressive versus learned are not always clear cut. For example, when Paul Sérusier, the symbolist painter best known for *The Talisman* (1888), refers to a synthesis of shapes into lines, angles, and arcs, Deleuze sees this is a reductive geometry; whereas he considers Paul Cézanne's geometric approach as not coded since it is about volume and, therefore, an analogical representation. Furthermore, even if painting is analogical because it is paralinguisitic or visually expressive, it is nonetheless digital because it is coded and learned, its expression based on historical conventions of style. And even though, as Deleuze reminds us, the digital aspect of art he is discussing does not involve mathematical convention but aesthetic, minimal units, some aspect of the digital are beyond code and convention.

Deleuze settles on a tripartite definition of code that hinges on its combinatory function in relation to art:

> One can make an intrinsic combination of abstract elements. One can also make a combination which will yield a "message" or a "narrative," that is, which will

> have an isomorphic relation to a referential set. Finally, one can code the extrinsic elements in such a way that they would be reproduced in an autonomous manner by the intrinsic elements of the code (in portraits produced by a computer, for instance, and in every instance where one could speak of "making a shorthand of figuration").[52]

Art can be a conventional application of coded elements to create abstract works in the manner of Herbin. It can also possess a pastiche function, an isomorphic relation to language or information. Or it can even present itself as an algorithm, a code, yielding an image of a recognizable figure. These combinatory aspects of code expressed by Deleuze in the context of art have their source in a more specific consideration of code that has repercussions in the very methodology of his multidisciplinary philosophy.

Code and the Markov chain

The combinatory function of code is explored further through chains of meaning in *Anti-Oedipus*. There, Deleuze and Guattari tackle the machinic code. In this conceptualization of code, the ideas of passage, connection, and transition are put forth and the chains of meaning operating through code seem to follow the virtual logic of the diagram.

The machine operates through code (Deleuze and Guattari start their example by focusing on a young boy whose thought is a machine; the code operating here is thus a Lacanian code of the unconscious, but, as we will see later, the code spills out of the unconscious and connects to the material). Within the code, a multiplicity of chains of meaning—made up of signs that do not signify—cut, break, and interrupt flows (a binary digital circuit comes to mind) between other flows or regimes or organs (of a Body without Organs). The connections are continuously reconfigured according to ever-changing input: "The data, the bits of information recorded, and their transmission from a grid of disjunctions of a type that differs from the previous connections."[53] How are the chains made up of signs that do not signify? The chains stand in for a general understanding of the term "code." Here, the analogy comes close to the theory of information insofar as we are dealing with information, and not words, that must enter into combinations that will generate meaning: "The code resembles not so much a language as a jargon, an open–ended, polyvocal formation."[54] This notion of jargon comes from a particular type of chain:

> Each chain captures fragments of other chains from which it "extracts" a surplus value, just as the orchid code "attracts" the figure of a wasp: both phenomena demonstrate the surplus value of a code. It is an entire system of shuntings along certain tracks, and of selections by lot, that bring about partially dependent, aleatory phenomena bearing a close resemblance to a Markov chain.[55]

The Markov chain is aleatory, or stochastic, and yet each step depends on the previous step in its chain of probability, which makes it random but transitional: each step is not determined by a fifty/fifty chance ratio but rather is informed by the condition of the previous state in the transition. The orchid and the wasp are heterogeneous but dependent.

Deleuze and Guattari's exposure to the Markov chain, as pointed out by the translators of *Anti-Oedipus* (and Anne Sauvagnargues below), came from Raymond Ruyer's *La genèse des formes vivantes*. There, Ruyer explains the notion of jargon, but also the possibility of applying the Markov chain's beyond coding to animal and biological milieus. In 1913, the Russian mathematician A.A. Markov devised the theory of probability bearing his name by scrutinizing Alexander Pushkin's *Eugene Onegin*, or, more specifically, by taking the first 20,000 letters (without spaces or punctuation) and making a long chain with them. He then determined how often vowels and consonants appeared sequentially.[56] As in Deleuze and Guattari's chains of meaning in which signs do not signify, the novel in verse was not examined for its signification but to see the patterns of occurrence of letters. As Brian Hayes explains, the text was treated "as a mere stream of letters,"[57] arranged in 200 grids of 10 × 10 letters. By calculating letter pairs Markov was able to determine that the probability of the occurrence of randomly selected letters in the sequence was not independent, but rather depended on a state of transition. Unlike the independent randomness of a coin flip, events in a system are connected: "Markov's methodology went beyond coin-flipping and dice-rolling situations (where each event is independent of all others) to chains of linked events (where what happens next depends on the current state of the system)."[58]

Following Markov's experiment with text, Ruyer provides an example of a machine (a conceptual machine of boxes holding assorted letters) that pastiches language based on a text by Livy. The probabilities of each random letter in this case are different from Pushkin's text since the language follows a different pattern and frequency of the letters appearing in Russian, Latin, or English.[59] The text produced reads as "IBUS CENT IPITIA VETIS IPSE CUM VIVIUS

SE ACETITI DEDENTUR," nonsensical jargon that nevertheless appears as Latin, a chain of signs that has no signification.[60] Ruyer writes: "The authors of pastiches instinctively conform to the typical frequencies of words or turns of phrases of their victims."[61] He applies this rule of pastiche to the animal world, shifting one milieu (text) to another (animal) while the function remains the same: "We notice that an animal could easily fall for a pastiche of the same genre taken from stimuli-signals that would interest it emanating from other animals of the same species."[62] He explains that animals signal to one another (here he gives the example of ostriches and wood pigeons) according to pattern that resembles a Markov chain: "a succession of semi-fortuitous themes evoked without a master plan and depending on the call of the preceding phase."[63] Animals can be taken in by hunters imitating their calls because here we do not speak of a language but of a jargon.[64] Ruyer applies the schema of the Markov chain beyond the animal world to many different milieus, including entomology, literature, philosophy, anthropology, art, and psychology. The last one is significant since it is tied to aesthetics and reveals the link to Deleuze and Guattari's use of the term. Ruyer writes that in less rigid domains, such as aesthetics, the relevance of chains of links resides in their disjunction. The example he provides is that of schizophrenia. Aesthetic forms, he writes, have disjunctions that are imprecise, as if somewhere between actual language and a dream language; in the case of humans, these disjunctions are apparent in the schizophrenic: "Associations become 'Markovian', they easily take an aesthetic hue."[65] Ruyer goes on to reinforce the aesthetic disjunction by comparing the aesthetic process of tincture in schizophrenic associations to the modifications that occur in the patterns of a butterfly's wings due to changes in temperature while it is in its cocoon. Not only is the terminology of jargon, chain, and schizophrenia interconnected between Ruyer's theories and Deleuze and Guattari's notion of code, but the function of partial dependence between phenomena is necessary for Deleuze and Guattari's own methodology of assemblage:

> In this sense it was possible to insist on a common characteristic of human cultures and of living species, as "Markov chains": aleatory phenomena that are partially dependent. In the genetic code as in the social codes, what is termed a signifying chain is more a jargon than a language *(langage)*, composed of nonsignifying elements that have a meaning or an effect of signification only in the large aggregates that they constitute through a linked drawing of elements, a partial dependence, and a superposition of relays.[66]

We can appreciate Deleuze and Guattari's diagrammatic application of the function of the Markov chain throughout different milieus—similar to Ruyer's implementation of this transduction in his writings.

Sauvagnargues makes the case for a coded approach to Deleuze and Guattari's nonlogocentric, multidisciplinary semiotics. She starts with a concept of a machine that is not technical nor technological but rather social. One that functions through a "combination (assemblage or *agencement*) of material elements."[67] The machine assembles and combines by cutting flows and coding. And a flow "is nothing other than another 'machine', captured as a matter of determination and not as an element of coding."[68] Sauvagnargues specifically aligns the notion of code with the Markov chain, whose stochastic process she defines as "the probabilities of occurrence of various future states [that] depend only on the present state of the system or on the immediately preceding state."[69] The Markov chain operates through states of transition. Code is not a language; rather, it is a "statistical 'jargon' of elements" operating according to "an iterative (automatic) procedure."[70] The Markov chain pushes past signification and symbolism into "statistical computation, just as applicable to life as to culture,"[71] because of its basis in information theory. How do we transition from language to computation to life to culture?[72] Deleuze and Guattari's use of transcoding between milieus appears to be an adaptation of Simondon's notion of transduction between matter (wasp/orchid).[73]

The Markov chain is the lubricant in the gears of Deleuze and Guattari's multidisciplinarity: "First, because it makes available a theory of order that is neither continuous nor determined; second, because the linking of signals is indifferent to signification as well as to the homogeneity of it element; and third, because it encompasses a theory of 'pastiche.'"[74] In fact, Sauvagnargues argues that Deleuze and Guattari's use of the Markovian code is essentially responsible for their moving away from a language-based semiotics to a multilevel rhizomatic system in which arrangements and assemblages between heterogeneous functions are possible, since it is the code that is transmissible from one milieu to the next and not the content:

> The Markovian code allows Deleuze and Guattari to break out of the enclosure of the human symbolic world and gain an entryway that allows for the unification of the new molecular biology and the discovery of the genetic code with economic theories and the cultural order in general, in a free enough variation that mixes political economy, the science of coding of flows, and especially, the Marxist theory of surplus value, freely interpreted as "surplus" and "capture of

> code." Operating between heterogeneous chains, it introduces a semi-aleatory order across disparate regions, which Deleuze and Guattari will soon describe as different "strata" of the material, of the vital and of the cultural.[75]

Sauvagnargues is able to make a direct link between the duality operating in the coded chain ("neither determinate nor fortuitous")[76] and the concept of the rhizome with its interdependent dualities inherent to its fundamental principles. Code, here, is at the basis of Deleuze and Guattari's methodology: rhizomatic organization proceeds through stochastic code, and this code opens up a spatialization, sometimes described as strata. In turn, an information theory based on spatialization of thought yields to an idea of a peripatetic and fungible map-image, since the map, in Deleuze and Guattari's theory, pinpointed the importance of the code in the first place.

One way of applying the Markov chain in the domain of art is by looking at the functional relationship between art and code. In the early days of computer art, Leslie Mezei literally applied transitional states captured in the form of a 3 × 3 grid inscribed with various degrees of probability to art. He used these mathematical models to create computer drawings by assigning values of probability to each transition state and then applying this matric to selected images. The example Mezei provides is his *O Canada* (1967), a computer drawing with a random distribution of various images associated with Expo 67 ("beaver, maple leaf, centennial symbol, Expo symbol"[77]). Another of Mezei's works, *Bikini Shifted* (1967), is drawn in lines that vary depending on the transitional values assigned to the figure of a woman, resulting in "a distortion effect not uncommon in some modern painting."[78] This is a literal application of code to art that Deleuze warned against in his book on Bacon, but here the function of code has the effect of modulating images, albeit not distorting them as Bacon does with his figures and faces. One could not compare a Bacon painting with a simplistic computer line drawing in terms of iconographic complexity, but the revelation here is that the code distorts rather than systematizes, as was feared. Whether it distorts systematically is, needless to say, the crux of the issue and the heart of the application of the Markov chain to algorithms.

But of course the Markov chain is not simply utilized to manipulate computer drawings. Its function has evolved since the 1960s to become an integral part of computing and analysis of complex patterns. Google, for example, uses it in their page ranking system.[79] Hayes explains how the PageRank algorithm is propelled by the Markov chain: "The transitions are links between pages. The aim of the algorithm is to calculate for each web page the probability that a reader following links at random will arrive at that page."[80] Gene identification, voice recognition, neurology, and physics are all domains that require the Markov chain. Kristen Daly enumerates some of the complex disciplines that have the stochastic process at the heart of their probabilistic systems:

> Complex functions govern computer software. Many tasks of the computer rely on discrete-state stochastic processes. These are nondeterministic processes, in that one state does not fully determine the next state. Complex patterns like neural networks and genetic algorithms require stochastic processes to map out processes. Unlike the ordinary differential equations of calculus, more than one possible outcome or state is conceivable in the evolution of the process. This indeterminacy is described by probability distributions. Therefore, from a given starting point or set of conditions there are a number of possible paths, some of which are more probable than others.[81]

And, indeed, the functionality of the Markov chain can be used as an aesthetic model, as Daly makes the case in the domain of cinema. This schema and application of probabilistic systems in different milieus converge for Daly in David Lynch's *Inland Empire* (2006), which, according to her analysis, functions through a process of assemblages of different data, like a chain of signs, that are given discreet signification based on adjacent states in the sequence of images. Lynch's self-reflexive film breaks down narrative expectations by disorienting the viewer, with scenes skipping across place (some take place in Los Angeles, others in Poland) and across time—even including surreal dimensions where prostitutes and rabbit-people seem to be part of a live-audience sitcom. The film, then, takes on the aesthetics of data processing of patterns in the form of disjointed narrative sequences of varying sources (from Lynch's hard drive), which functions as—to use Ruyer's phrase in applying the Markov chain to art—a dream language.[82]

We can see how much closer we are to the code emerging from the rhizomatic map/tracing duality. The combinatory function of the Markov chain as suggested by Deleuze in his consideration of code in painting and then with Guattari in *Anti-Oedipus* as a philosophical method yields to a far-reaching aesthetic code that permeates painting, film, as well as a multitude of technological cultural objects. It is through the concept of the map that we can approach code as a cultural, visual object.

Mapping digital poetics

Even though mapping spawned the concept to the code in Deleuze and Guattari's aesthetics, we have to get back to mapping *and* coding. How does mapping fit into an aesthetic of code? Jussi Parikka suggests a strategy to bridge mapping and the digital in his "Ethologies of Software Art" by focusing on the "imperceptible" aspect of digital art. The imperceptible, according to Parikka, is supported by concepts such as "affects, sensations, relations and forces," and these are "nonhuman" insofar as they "exceed the modes of organisation and recognition of the human being."[83] Parikka's imperceptible carries dualistic yet dynamic interrelations, namely, the incorporeal and material—or, as Sauvagnargues puts it, the combination of social machines and material elements.[84] This concept is in line with the way that code in language transpired into code in biology in Ruyer's theory. In this case, the incorporeal is seen as "the ephemeral nature of the event as a temporal unfolding instead of a stable spatial identity"; whereas the material is defined as "an intensive differentiation that stems from the virtual principle of the creativity of matter."[85] Both the incorporeal and material carry within them this notion of the imperceptible "as a futurity that escapes recognition."[86] Parikka relates the code in digital art to the process of translation: "Software is not (solely) visual and representational, but works through a logic of translation. But what is translated (or transposed) is not content, but intensities, information that individuates and in-forms agency; software is a translation between the (potentially) visual interface, the source code and the machinic processes at the core of any computer."[87] Here, we can see how translation between code and representation echoes the Markovian concept of chains.

This is relevant from an intermediality point of view, which is at the basis of digital art, a process we have seen in the relationality between tracing and mapping and the transitional part of the Markov chain. What we are looking for is not the technical aspects but the "poetics of potentiality."[88] Coding becomes a lens reflective of culture: "thinking of code not only as the stratification of reality and of its molecular tendencies but as an ethological experimentation with the order-words that execute and command."[89] What Parikka is advocating for is the application of ethology—knowledge dealing with the formation and evolution of human ethos. It is essentially the science of human character or sentiment, and in this case the human character is formed and modeled through algorithmic thinking. Beyond the cybernetics use of the human, here we assess its cultural effects. Thus, the hierarchization of the algorithmic as decried by

Rosenstiehl and Petitot resurfaces. The question of the imperceptible fits in with the notion code since it is not readily available for scrutiny yet its effects are quite real. The imperceptibility of code is something both aesthetic and political: "Code is imperceptible in the phenomenological sense of evading the human sensorium, but also in the political and economic sense of being guarded against the end user."[90] The code, imperceptible yet active, seems on an aesthetic level to enact the virtual/actual duality. This is how Parikka grafts the digital onto Deleuze's theories:

> Imperceptibility should not be understood solely in terms of the representational logic of what is seen and what is not. It refers as much to the powers of virtuality as well. In terms of approaching the virtual element of reality as one of potentiality not exhaustible by any (already) realized and extended actualities, we can appreciate the idea of this core of creativity as inherent in both a Deleuzian ontology and a digital ethology.[91]

Imperceptibility as the virtual is a guiding aesthetic siphon through which to filter the expanse of the digital into a Deleuzean philosophy. In that case, we would see mapping as virtual and tracing as actual.

Parikka uses the word ethology to underline the overflow of meaning of the digital as it relates to art and culture. Digital is not simply zeros and ones but a whole other strata of culture and understanding: namely relationality, an understanding of materiality that touches upon the noncorporeal, and a whole rhizomatic netting of culture at large, as well as the intermedialities of digital interfaces.[92] Bringing things back to Deleuze, Parikka reminds us of the tense relationship the philosopher had with his own aesthetic interpretation of the concept of the digital and the origin of his negative view of the term: "the notion of code as a formalisation and standardisation of the rationality of the analogue."[93] The calcification around categories of digital and analog is a problem that resonates in either category. Softening the duality by acknowledging cross-pollination between seemingly opposite categories could reinstate the poetic flexibility for code within the realm of aesthetics, as we will see shortly.

Parikka sees code as something that is already part of culture at large, and he brackets out the deeper philosophical descriptions of how the virtual operates in Deleuze's philosophy. He is careful not to use the term, but rather engages with the imperceptible, an incorporeal agency that is manifest in actuality. And perhaps this is our way out of the technical complexity of algorithmic processes. The forced compatibility between Deleuze's notion of code/digital and contemporary usage of these devices could be limiting. Understanding the notion of coding in an ethological sense, or a functional aesthetics, especially one that is applied to an object like art, is the level at which the present study wishes to operate. Parikka offers an example of the digital wrapped in a sheath of ethology, operating directly within an artistic aesthetic that relies less on art objects and more on networks:

> The relationality of ethological assemblages provides a potentiality that several software art projects aim to tap into. They do so in order to map lines of becoming by injecting themselves into a logic of networks, software and the wider media ecology of contemporary culture. By functioning according to the same logic utilised by consumer software in control society, software art aims to find the cracks in the majoritarian language operating as the cultural relay of power and control.[94]

Software art (here we can think of etoy or Jody.org, with its pastiche/subversion mechanism) functions precisely as tracing and mapping, subverting the machinic through an exact mapping on its own territory.

Code, therefore, in its aesthetic, poetic iteration still has the possibility of entering into the cultural domain of communication through semblance and representation. But, as we will see with Alexander Galloway, this representation of code and information is often elusive and hard to come by.

Data aesthetics

What is an aesthetic approach to information? The aesthetics of information, data, and their relationship to the communicable through representational means is at the heart of Galloway's *The Interface Effect*: "Is data meant to be textual and static, while information elastically structured via flows and arrangements?"[95] Imperative for us here is how data can be visually represented. Similar to Parikka and his poetics of potentiality, Galloway calls for a "poetics of the algorithmic." This is problematic since Galloway tells us that "data have no necessary visual form."[96] What becomes apparent in his example of the maps of the Internet is that the visualization of something incorporeal like data is always filtered through an illustrator's set of aesthetic considerations: "Data, reduced to their purest form of mathematical values, exist first and foremost as number, and, as number, data's primary mode of existence is not a visual one."[97] But Galloway strikes an optimistic note about the representation of data sets, since the solution for its poetics resides in coding: "This does not mean that aestheticization cannot be achieved. And it does not mean that such acts of aestheticization are unmotivated, nugatory, arbitrary, or otherwise unimportant. It simply means that any visualization of data must invent an artificial set of translation rules that convert abstract number to semiotic sign."[98]

The problem is oddly similar to that outlined by Deleuze in the codification of abstract painting. But what we are doing here is a contrary movement: starting with information, going through code, and getting to the aesthetic visual object. Since there is no formal way of depicting data, Galloway turns to Deleuze's concept of the virtual to illustrate, through philosophical means, the impossibility of purely representational information: "Thus in Deleuze information isn't the bubbling chaotic material plane, but rather what Deleuze calls the virtual."[99] One of the solutions Galloway proposes is that of network visualization: "For just as network visualization can tend to obfuscate its own data, it may also reveal systems of organization and power, given the right conditions."[100] Here Galloway has in mind Mark Lombardi's charts or Bureau d'études' diagrams. Lombardi's circuitous charts map out global networks of corruption involving politicians and corporations. Bureau d'études utilizes visualizations and diagrammatic display to elaborate on the dark corners of politics.[101] In fact, mapping is the pivot point between data and information from the perspective of aesthetics because "the ideological content of the map is ultimately beholden to the affordances and prohibitions of form" and this is ultimately manipulable.[102] Instead, Galloway warns, "let us not tarry with the various attempts to critique the social map at the level of data, and instead consider some of the attempts to critique it at the level of information."[103] The murky side of diagrams is something we will be dealing with, but it is worth noting that Galloway sees the ideological manipulation of seemingly objective visual devices at the level of the difference between data and information. And it is at this level that he calls for a "poetics for such algorithmic systems."[104] The aesthetics of coding emerges as one that stands in between the virtual, imperceptible dimension of culture and the way that it is made actual, representational. This space between these two instances is what needs to be charted through the map-image.

Perhaps a note on Bruno Latour's clarifications of what consists a network would be helpful in the context of charting information aesthetically. In his Actor-Network Theory (ANT)—in which society (composed not simply of peoples but of atoms, trees, and books) is gauged according to actions perpetrated by actors (defined as those who perpetrate actions) and therefore is seen not according to old hierarchical categories but rather on a leveled playing field—the network is hard to imagine. Latour concedes that a computer network, with its complex connections and its high level of organization, might be a perfect image of the network he has in mind, yet it is not.[105] The network is understood not in terms of surfaces, but instead as a constellation of adjectives that refer to Deleuze and Guattari's rhizome: a "fibrous, thread-like, wiry, stringy, ropy, capillary character that is never captured by the notions of levels, layers, territories, spheres, categories, structures, systems."[106] Latour also reverts to geographical terminology to provide a visual representation of what he means by the point of focus of his theory: "Loci, contingencies or clusters are more like archipelagos on a sea than like lakes dotting a solid land."[107] Latour, on the problem of representing the notion of network spatially, a notion that admittedly is easily deployed spatially in the visual imaginary, wants us to focus on the connections and not the typical landmarks: "The notion of network helps us to lift the tyranny of geographers in defining space and offers us a notion which is neither social nor 'real' space, but associations."[108] Of course, this being a book about the map-image, it still accommodates Latour's statement, since the map-image is a way of understanding space not through an expected cartographic cliché but, as I announced at the beginning of these pages, as a fluctuating image of a map and not the map itself.

The difficulty of representing a map of the informational network is at the center of Galloway's concern about "the nature of information aesthetics."[109] And his question regarding the representability of data or information relies on Jacques Rancière's questions around what is unrepresentable. Rancière surveys the impossibility of representing violence in theater, poetry, literature, and film according to each medium's own formalistic rules on representation. He settles on a Lyotardian version of the sublime, which itself is inspired by Hegel. Rancière turns to Abstract Expressionism as an example of the sublime that captures the unthinkable "through the orange-coloured flash of lightning that traverses the monochrome of a canvas by Barnett Newman, or any other procedure whereby painting carries out an exploration of its materialism when they are diverted from the task of representation."[110] The unthinkable (in this case, the Holocaust) is difficult to adequately represent artistically. The excess of information of such a historical event and its formal capture lead to an incongruity Rancière summarizes as "a concept of art and a concept of what exceeds art."[111] Hegel, in his aesthetic categories—symbolic, classic, and romantic—demonstrated how form and content struggle to come into harmony. If form and content are adequate in classical representation, this is short-lived. The excess of art creates a sort of *aufhebung* in the romantic—art is divided into three forms: painting, music, and poetry, themselves pointing to a whole other form of expression, philosophy. For Galloway, however, who takes up Rancière's notion of unrepresentability and applies it to networks, it is not a question of content but of the apparatus that is responsible for the content. The question, according to Galloway, is not, as Rancière asks, whether representations of violence affect us, but why are we blind to the apparatus, the mode of production of the society of control, which has not yet been put into view.

These questions of unrepresentability, of the end of art, of imperceptibility, can be folded into the aesthetic category of virtuality, and especially the way it operates in digital aesthetics.

Virtuality of digital art

How is the virtual an aesthetic image of the digital with all its visual and imaginary permutations? Associated with this notion of coding is an aestheticization of digital art through the notion of the virtual and actual. Laura U. Marks argues for a poetics of data and algorithm when she explains the relationship between the image and the algorithm in digital art: "In computer art, the image is the mere skin of an artwork whose underlying structure and raison d'être lie elsewhere: in its algorithm and database."[112] She defines the algorithm's role in the production of the image in digital art as follows: "Algorithmic structure: a structure based on a series of instructions that manipulate information to produce actions or images."[113] Ultimately, the logic between the algorithm and the image is that of folding and unfolding. She compares a computer art piece by John F. Simon Jr. in which square patterns appear to open up like box lids seemingly to infinity with the click of a mouse with the Sultan Hassan mihrab, the highly decorated fourteenth-century Egyptian mosque, stating that both these aniconic, abstract, yet highly coded works function around the notion of enfoldment: "The two works of art evoke the difference between a fathomable infinity—that encoded in Simon's software—and an unfathomable infinity, that of the Deity to whose unknowable presence the mihrab points."[114] Marks also draws the translation of mathematician Mohammad Ibn Musa Al-Khwarizmi's name into the Latin word *algorithm* as a series of translational foldings, which begin with *algebra* rooted in the Arabic *al-jabr*, or restoration, in twelfth-century Toledo.[115] This revelation, translation, and enfolding are ways of using the virtual as a means of explaining the relationship between the image and the algorithmic code: "What is unfolded into information or image can be considered actual, while what remains enfolded remains virtual."[116] Deleuze's fold is paired up with his notion of the virtual. It is in his treatment of the cinematic image that the virtual is applied to a particular type of self-reflexive mirror image, which, while not exactly directly manifested in Simon's algorithmic works, nevertheless demonstrates the aesthetic application of this complex philosophical concept. We will explore this further in Chapter 3.

The aesthetics of code (ethology of imperceptibility; visualization of informational networks; undermined geographical visualization of spatialized abstract networks; the *aufhebung* of the unrepresentable; enfolded virtuality) in its many iterations of strategies captured through visual representation of algorithmic culture circles around and back to mapping, tightening the deleuzoguattarian cartographic dependence on code. But how does this play out in actual instances of art that uses mapping devices to express a culture based on information, data, and code? This can be witnessed through the confrontation of mapping and informational images in art history. After all, as I stated before, art is instrumental in articulating how code weaves itself through different aesthetic milieus.

Aesthetics of documentation in Savard and Lagrange Paquet

It is clear by now that in discussing the map-image we are not dealing with representational, illusionistic, or figurative art, but instead with art that utilizes images of code, maps, and information. In the domain of art, we can see informational images in the context of traditional media, such as painting, and new media, such as algorithmic art. The way code is used in these two media is illustrated by the painting of Francine Savard and the algorithmic art based on video games of Emmanuel Lagrange Paquet.

But first, let us consider how art history treats informational images. James Elkins, in his "Art History and Images That Are Not Art," deals with informational images. He starts with a colorful analogy. He writes: "An image taken at random is more likely to be an ideographic script, a petroglyph, or a stock market chart than a painting by Degas or Rembrandt, just as an animal is more likely to be a bacterium or a beetle than a lion or a person."[117] These images that are not art do not have a proper name to classify them. That is one of the reasons Elkins's title defines them negatively: by what they are not. But he explains that this "group" consists of "images principally intended—in the dry language of communication theory—to convey information." These half-pictures, hobbled images, are "bound by the necessity of performing some utilitarian function," hence their association with text and numbers. This nameless group of pictures includes "graphs, charts, maps, geometric configurations, notations, plans, official documents, some money, bonds, seals and stamps, astronomical and astrological charts, technical and engineering drawings, scientific images of all sorts, schemata, and pictographic or ideographic elements in writing."[118] I want to briefly examine the incursion of these types of images into the realm of art. Elkins believes that the relationship between these images and art history is crucial since the discipline of art history has developed a language about pictures and their understanding.[119] Nevertheless, there is a fundamental difference between informational images and art images insofar as the former need a degree of specialization that is not present in the interpretation of images that are meant to communicate expressively: "Nonart images are even less stable, because they depend less on resemblance and more on specialized interpretive skills that are easily shifted and lost over time."[120] Thus, when dealing with the incursion of informational images into art historical categories (painting, collage, printmaking), these images have to be looked at with their documentary sources in mind.

I will look at two instances of informational images by two Montreal-based artists. In the first, I will investigate how Francine Savard deals with mapping and coding through painting. We will go back to Steinberg's orientation of the picture plane as a repository for information, as it serves as a ground to "read" Savard's works in the vein of conceptual art. I will then compare Savard's use of mapping in painting to mapping in the digital field. For this purpose, I will analyze Emmanuel Lagrange Paquet's translation of video games into art and the documentation of a territory consisting of action and space that could best be described as haptic.

Savard labors at the edge of the self-reflexive plane of art with works consisting of indeterminate maps and tables of data. For example, Savard's map *Dépot de peinture* (2000) [*Figure 1*] presents not figurative objects, but nonfigurative images that bring attention to a conceptual process of conveying information about the very painting that contains these visual objects. The image here is *like*

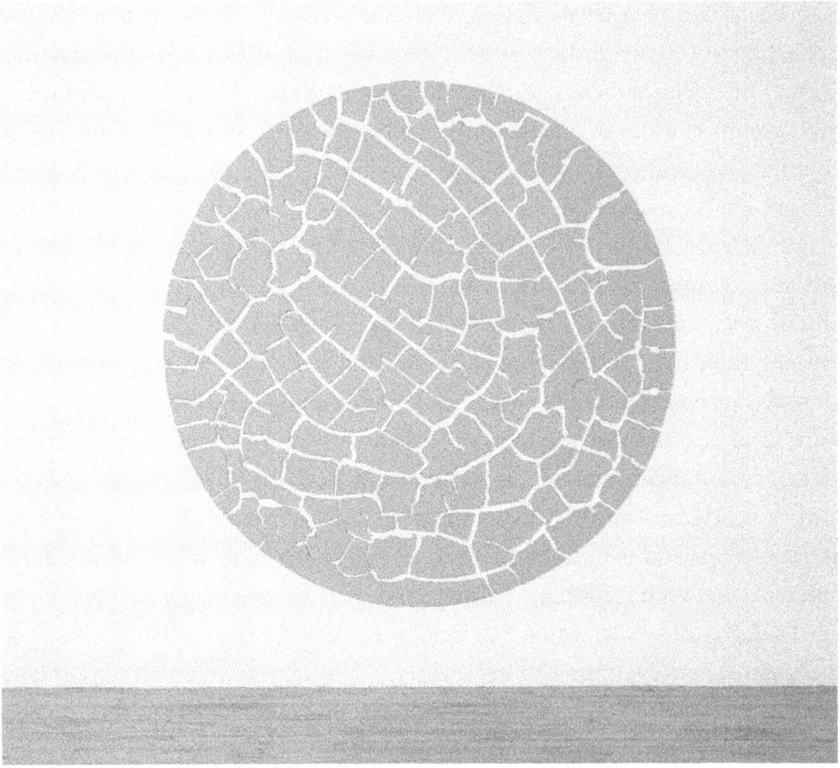

Figure 1 Francine Savard, *Dépôt de peinture* (2000). 125 éléments. Acrylique sur contreplaqué russe. Diamètre: 240 cm, Collection Musée d'art contemporain de Montréal Photo: François LeClair. Courtesy of the artist.

an image in the same way that Kant described his schema as an image that is not quite like an image. It is also an ironic play on abstraction in painting, yielding information even when there is nothing to communicate. The cartographic element of the diagram cements the concept within the boundaries of space. The map is a metaphor for the potentiality at work in the artistic process.

In a similarly ironic fashion, Savard presents a work of visual mapping—epigrams on a table, a flatbed picture plane, as it were—giving it the name *tableau*, as in, painting. *Tableau chronologique* (2009) summarizes the arrangement and intricate connective details between literary works, citing in epigrammatic form authors from the past: an informational image, clearly in Elkins's sense of the term, with its connection to text. The formal aspect of the table of information here reacts to time, as past authors are more likely to be cited by more contemporary ones. This results in a frenzied network of lines toward one end of the picture plane, clearly demonstrating the passage of time in the way that Charles Joseph Minard's famous *Carte figurative des pertes successives en hommes de l'armée française dans la campagne de Russie, 1812–1813* (1869) does, but in far more streamlined and abstract fashion—and also on a far broader historical expanse. For Savard, the subject matter is information and the shape of its presentation.

How can we then summarize the directionality of abstract painting resembling a map and a flat table that qualifies as a tableau? Steinberg's concept of the flatbed picture plane is instructive here. The term comes from the flatbed printing press, or receptor, "on which information may be received."[121] The flatbed picture plane changes the nature of the content of a painting through a spatial reorientation. This tilt of the picture plane from vertical to horizontal is expressive of the most radical change in the subject matter of art: it is the shift from nature to culture.

Referring back to Steinberg reminds us that if the vertical picture plane is associated with figuration, human posture, and nature in the works of Renaissance and Abstract Expressionist artists alike, then the flatbed picture plane brings forth a new era of representation with its horizontal orientation. The flatbed picture plane is equated with surfaces like "tabletops, studio floors, charts, bulletin boards—any receptor surface on which objects are scattered, on which data is entered, on which information may be received, printed, impressed—whether coherently or in confusion."[122] The term's printing press origins remind us that information that was transmitted started off horizontally. A radical and new orientation takes place: the picture plane does not refer to a visual experience one finds in nature, but rather to a conceptual interaction with "operational processes." We are not seeing something illusory but rather informational.

The informational relationship between painting and printing press is quite obvious in Savard's suite of paintings dealing with weather forecasts that she collected over the course of a year from the front page of *Le Devoir*, a daily newspaper in Montreal. This suite comprises box-like monochromatic canvasses painted in solid colors with the words painted in a different color grade in the center. Those words are part of the title of the works: for example, the words *Pluie intermittente, venteux, neige fondante* are at the center of the paintings entitled, respectively *Pluie intermittente* (*Précipitations 4*) (2013); *Venteux* (*Élements 3*) (2013); and *Neige fondante* (*Précipitations 4*) (2013) [*Figure 2*]. The weather conditions that make up the paintings (which are then classified in parentheses according to their nature as element or precipitation) are captured in a paper

Figure 2 Francine Savard, *Neige fondante (Précipitations 4)*, 2013. Acrylique sur toile marouflée sur structure de contreplaqué. 37.5 × 49.5 × 24.8 cm (14 ¾″ × 19 ½″ × 9 ¾″), Collection privée, Toronto Photo: Guy L'Heureux. Courtesy of the artist.

copy, printed black on white, with the information that came out of the newspaper weather section but in one seemingly singular sentence: the document has the appearance of data flow, like the data-rain from the matrix or Markov's letter grids. *Le temps qu'il fit (à la une du "Devoir")* (2013) is like the master code from which the information was parceled into individual boxed canvasses and designated with individual color [*Figure 3*]. This particular piece is horizontal, like the flatbed of a newspaper. Too dense to read, with redundant information, and monotonous in its subject, the work brings out the aesthetic yet posthuman aspect of data. The works in their totality are produced through a code, like the one Deleuze wrote about in relation to Auguste Herbin and his color code of language and form. This coding process is summarized by Sarah Milloy:

> Taking as her starting point the short-form weather forecasts from the front page of her Montreal newspaper over a period of year (think words and phrases like "sunny," "cloudy," and "snow flurries"), she then analyzed her findings, making

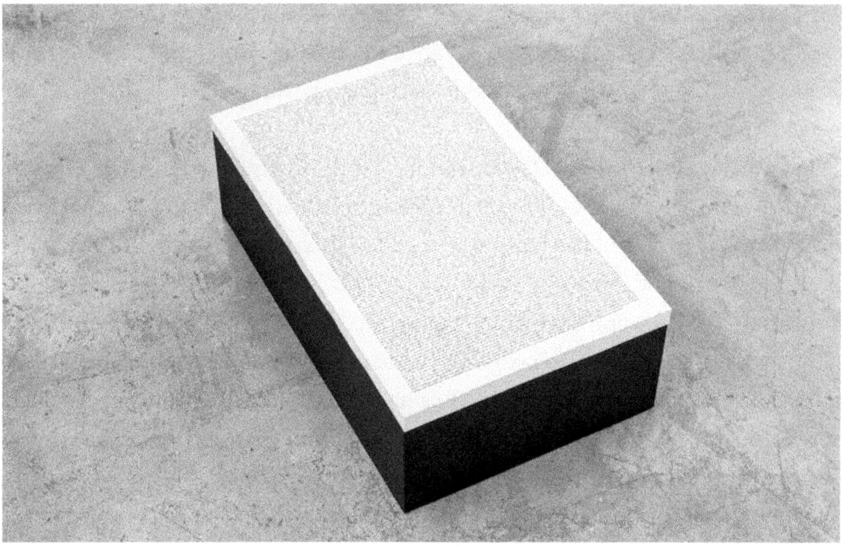

Figure 3 Francine Savard, exhibition view *Weather* (2013), Diaz Contemporary, Toronto Photos: Toni Hafkenscheid. Courtesy of Diaz Contemporary, Toronto.

> note of the frequencies to produce a suite of objects of variable size—canvas stretched over rectangular frames and painted front and side—each of which bears the word or phrase in question and the colour that Savard associated with that word. At the centre of each canvas, the term itself is rendered in small text, subtly embedded in the monochrome field like a language mirage.[123]

The process of finding and compiling data, observing a transcription procedure, and visualizing the data in minimalist-like canvasses seems to show a different type of landscape, one that is interested in information and not in actually looking out the window. As Steinberg summarized it: "This is not the world of the Renaissance man who looked for his weather clues out of the window but the world of men who turn knobs to hear a taped message, 'precipitation probability ten percent tonight,' electronically transmitted from some windowless booth."[124]

But if the shift for Savard was made from vertical to horizontal, from wall to table, for Lagrange Paquet, the horizontal is literally parallel with the earth's horizon, and we are no longer above a table but much farther above, in fact nullifying human posture altogether. Savard takes her coding from newspaper flatbed sheets of information; whereas with Lagrange Paquet, we are moving away from the information age into the codification of the digital age in which the canvas turns to screen. Lagrange Paquet's is a world of algorithms; yet we find, as Betsky advocated, a resistance to code, or a manipulation of code through the digits of the hand. Code here moves from screen to paper, from eye to hand.

With his series *Histoires d'interactions* (2014), of which *Boston Tea Party* is a part [*Figure 4*], Lagrange Paquet provides information images of experience vis-à-vis the interface of the video game, although this information is not always clear. He wants the representational experience to be steered away from the three-dimensional world into which players inject themselves and rather remain at the surface of the interface experience, which are the buttons on the controller.[125] He establishes a grid that can be read in a linear fashion in order to decode which buttons on the controller were pushed at which frequency and in which order. The order is revealed through the triangle, square, circle, and X, as well as the direction of the controller. The patterns and the symbols can be recognized by players, but as a result of the printout's presentation in the gallery space, we don't know which moments in the virtual space the sign sequence corresponds to.[126]

Lagrange Paquet's work also has a documentary, archival aspect. He writes:

> Even though historians do better work than those of Ancient Greece, rare are the people who have really participated in such events as the storming of Normandie. What we know about it, what has been transmitted, told or shown through fragments of photographic archives is the subject of, in a distant future, a process of mythification and plays on the distinction between fiction and reality.[127]

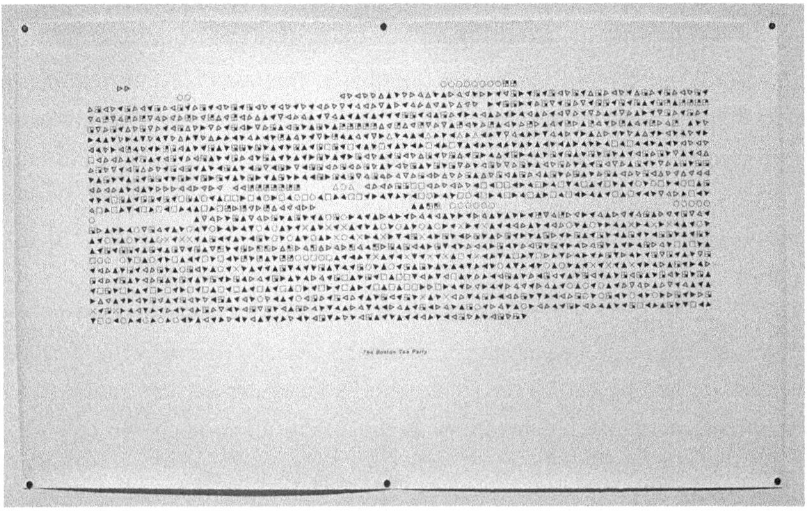

Figure 4 Emmanuel Lagrange Paquet, *Histoires d'interactions* (2014). Courtesy of the artist.

Lagrange Paquet operates somewhere between fiction and documentation and his symbols appear as runes (to use Craig Owen's term) even when it is understood they are meant as graphs expressing the sequence of a controller's buttons being pressed: like an ancient language that we cannot decipher of a mythical battle that we cannot see. The signs, or runes, are presented in two formats that, according to Lagrange Paquet, have an association with the documentary aspect of information: the portrait format, which presents the runes as if they were an instant, objective photographic rendering; and the landscape, which hints at a historical official representation of a battle. But these signs have a specific origin; they are more graphic traces of the action of the body.

What do we do when the body's interaction with the screen of information is the game's primary function? Pasi Väliaho explores the space of interaction between a viewer and the virtual environment of a video game's operational area. The level of engagement is rather complex, as Väliaho explains: "Players perceive their game environment as a stream of constantly changing rhythmic stimuli that elicit their motor behavior. Indeed, their perceptions cannot be separated from moving about, probing, and interacting with their virtual surroundings."[128] The relationship between the player and the action on the screen is mediated by a series of hair-trigger responses corresponding to the sequences of buttons pressed by the thumb on the controller. Lagrange Paquet's work enshrines the indexical trace resulting from movement of the body without the corresponding visuals of the virtual world into which the digital apparatus translates the gestures of the body. Accordingly, the inscriptions of the gestures, or graphic recording of the body's relationship to the shifting image on the screen, are indexed, framed, and remain elusive.

In her essay "Index, Diagram, Graphic Trace," Margaret Iversen distinguishes between three types of signs. She writes that "the index has a close, causal or tactile connection with the object it signifies." The diagram is involved in a coding or translation exercise, representing incorporeal events visually, such as statistics originating in the stock market or the weather. She explains that some diagrams are "generated directly from their object." This type of diagram, which she calls a graphic trace, belongs to its own hybrid category—"it takes from the index a registration of something unique" and incorporates the abstract quality of the diagram. Iversen seems to speak directly to the issues of representation Lagrange Paquet puts forth through his works: "The graphic trace is a diagram actually generated by the body and as such it combines the carnal and the

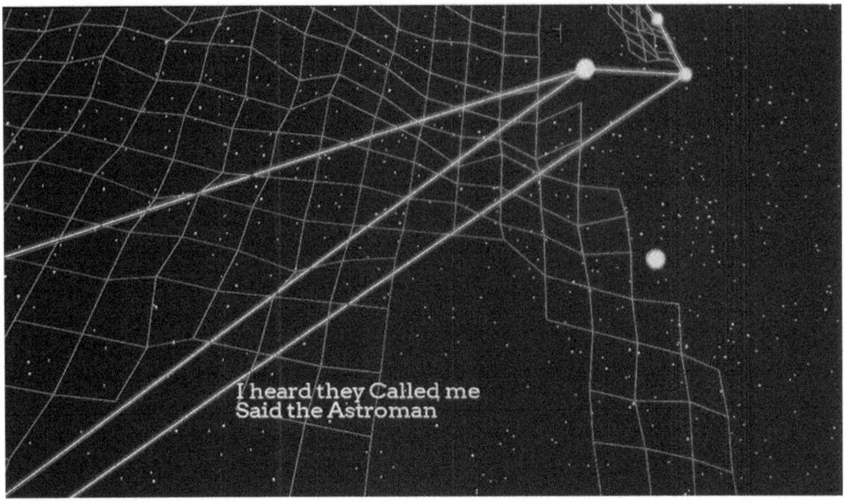

Figure 5 Emmanuel Lagrange Paquet, *Les cosmos imaginaires* (2014). Courtesy of the artist.

symbolic, the line and the flesh, effortlessly, automatically."[129] In this exchange of information through different codes of varying levels of conventional decoding we find the mapping aspect between the tactile gesture and digital visualization in the work *Les cosmos imaginaires* (2014) [*Figure 5*]. This work consists of a screen projecting a display resembling the cosmos (speckled with stars and gaseous colors) on top of which are displayed sentences and graphs (lines, grids, moving, bending). The words incorporate the lyrics that touch on the theme of space. The graphs are made and manipulated by the hand of the viewer through a Kinect camera, following an algorithm (cv.jit.features) that allows them to create a personalized map of constellations and the resulting poetic lyrics and sounds. *Les cosmos imaginaires* uses coding as a way of creating new ways of approaching, mixing, and experiencing culture. Perhaps cartography—when taken within the fold of coding and art, becoming a mixture of potentiality and representationality, self-reflexivity and the conceptual—can be a productive lens through which to observe informational images in the context of art.

Conclusion

Digital aesthetics are an ethology (Parikka) and a politics (Galloway) because they do not have a clear ontological source in Deleuze. Digital aesthetics are virtual, imperceptible, and unrepresentational. The map-image is the virtual source of a Deleuzean digital aesthetic. The concept of the map becomes a hodology of the digital. Starting with the maps within Deleuze's (and Guattari's) multidisciplinary aesthetics—such as Bacon and the geography of painting, Resnais and the visual cartography of time, and Kafka's system, itself treated as a map with multiple entrances and exits—we have revealed an equally significant stock of key terms related to informational aesthetics within these sources. The geography of Bacon's painting is contrapuntal to a coded (or overcoded) space on the canvas. As for the network arrangement of a spatialized reading of literary corpus, it is only the beginning of a rhizomatic cartography expanded in *A Thousand Plateaus*. The treatment of the map in conjunction with tracing within the concept of the rhizome elaborates more firmly not only the aesthetic dimension of the map but also its relationship to abstract space. More concretely, though, it is there that we are introduced to the coded dimension of the map itself. Through chart theory in algorithmic thinking, Deleuze and Guattari hint at an informational application of the map. Mapping helps capture the diagrammatic exchange between milieus. The Markov chain, the source of the code aesthetic for Deleuze and Guattari, informs the mapping concept and is also part of their methodology. The philosophical concept of the map-image, in its shuttling between abstraction and representation, navigates between intermedial artistic manifestations of information from painting, where it is represented, to multimedia installations, where it is both subject and process. The map-image, amorphous and self-reflexive, weaves itself through philosophy and art.

Notes

1 See Deleuze and Guattari, *What Is Philosophy?*, 191.

2 Ibid., 15–16.

3 Steinberg, *Other Criteria*, 84.

4 Deleuze, *Cinema 2*, 313n11.

5 Deleuze, *The Fold*, 146n2.

6 Deleuze and Guattari, *What Is Philosophy?*, 44.

7 Deleuze, *Essays Critical and Clinical*, 191n13.

8 Deleuze, *Cinema 2*, 122.

9 Ronald Bogue, *Deleuze on Literature* (New York: Routledge, 2003), 173–74. Bogue considers the artistic interventions on Voie Suisse by Carmen Perrin and others and their relationship to virtual paths in an actual landscape.

10 Webster's Dictionary.
11 "It could be said that the orchid imitates the wasp": Gilles Deleuze and Félix Guattari, *A Thousand Plateaus: Capitalism and Schizophrenia II*, trans. Brian Massumi (Minneapolis: University of Minnesota Press, 2005), 10.
12 Ibid.
13 Ibid., 12.
14 Ibid.
15 Ibid.
16 Deleuze and Guattari, *Kafka*, 3.
17 Ibid., 37.
18 Deleuze and Guattari, *Thousand Plateaus*, 12.
19 Deleuze and Guattari, *Kafka*, 10.
20 Deleuze, *Francis Bacon*, 184n3.
21 Ibid., 90–91.
22 Deleuze and Guattari, *Thousand Plateaus*, 13.
23 Ibid.
24 Ibid.
25 Ibid., 14.
26 Ibid., 16.
27 Ibid., 17.
28 Ibid., 21.
29 Ibid., 20.
30 Ibid., 21.
31 Pierre Rosenstiehl and Jean Petitot, "Automate asocial et systèmes acentrés," *Communications* 22, no. 1 (1974): 49.
32 Ibid., 50.
33 Ibid., 61. "Nous avons résolu des problèmes de coordination entre cellules sans coordinateurs en nous octroyant l'hypothèse forte de la calculabilité, c'est-à-dire de la digitalisation de l'information. En d'autres termes, la nature des cellules, et les problèmes à résoudre ont toujours été réduits à des listes d'états susceptibles d'être représentés par des numéros (digits) et à des règles de changement d'état également digitalisables."
34 Deleuze and Guattari, *Thousand Plateaus*, 313.
35 Ibid.
36 Ibid.
37 Alexander Styhre, "The Concept of Transduction and Its Use in Organization Studies," *E:CO* 12, no. 3 (2010): 120–21, quoting Steven Shaviro: Simondon on Individuation: www.shaviro.com/Blog/?p=471.
38 Deleuze and Guattari, *Thousand Plateaus*, 314.
39 Young, Genosko, and Watson, *Deleuze and Guattari Dictionary*, 68.

40 Ibid., 141.

41 Ibid., 176.

42 Anne Sauvagnargues, "Desiring Machines and Social Coding," in *Artmachines: Deleuze, Guattari, Simondon*, trans. Suzanne Verderber with Eugene W. Holland (Edinburgh: Edinburgh University Press, 2016), 210.

43 Virtual: "Boundas describes the virtual as consisting of incorporeal events and singularities, and this perhaps is a new way of thinking software art as well: they summon such singularities both as embedded in the nonhuman materiality of computers, and in the incorporeal events acting as experimentations." See Jussi Parikka, "Ethologies of Software Art: What Can a Digital Body of Code Do?" in *Deleuze and Contemporary Art*, ed. Stephen Zepke and Simon O'Sullivan (Edinburgh: Edinburgh University Press, 2010), 119–20.

44 Deleuze, *Francis Bacon*, 104.

45 Ibid.

46 Ibid.

47 Ibid.

48 Ibid.

49 Ibid.

50 Ibid., 109.

51 Ibid., 112.

52 Ibid., 114–15.

53 Gilles Deleuze and Félix Guattari, *Anti-Oedipus: Capitalism and Schizophrenia*, trans. Robert Hurley, Mark Seem, and Helen R. Lane (Minneapolis: University of Minnesota Press, 2000), 38.

54 Ibid.

55 Ibid., 39.

56 Brian Hayes, "First Links in the Markov Chain," *American Scientist* 101 (2013): 95.

57 Ibid.

58 Ibid., 93.

59 Raymond Ruyer, *La genèse des formes vivantes* (Paris: Flammarion, 1958), 172.

60 Ibid.

61 Ibid. (my translation).

62 Ibid. (my translation).

63 Ibid. (my translation).

64 Ibid., 173.

65 Ibid., 186–87 (my translation).

66 Deleuze and Guattari, *Anti-Oedipus*, 289.

67 Sauvagnargues, "Desiring Machines," 206.

68 Ibid.

69 Merriam Webster's Dictionary.

70 Sauvagnargues, "Desiring Machines," 207.
71 Ibid.
72 Ibid.
73 Ibid. "For Ruyer, Markov chains thus determine the mode by which forms communicate, a mode that is non-final but also not lacking in order: in this way, an animal allows itself to be taken in by pastiches of the stimuli-signals that it finds interesting, something hunters know very well."
74 Ibid., 207–8.
75 Ibid., 209.
76 Ibid., 210.
77 Leslie Mezei, "Randomness in Computer Graphics," in *Cybernetics, Art and Ideas*, ed. Jasia Reichardt (Greenwich: New York Graphic Society Ltd., 1971), 168.
78 Ibid., 171.
79 Kristen Daly, "Cinema 3.0: The Interactive-Image," *Cinema Journal* 50, no. 1 (2010): 95.
80 Hayes, "First Links," 92.
81 Daly, "Cinema 3.0," 94–95.
82 Ibid., 95.
83 Parikka, "Ethologies of Software Art," 116.
84 Sauvagnargues, "Desiring Machines," 206.
85 Ibid.
86 Ibid.
87 Ibid.
88 Parikka, "Ethologies of Software Art," via Matthew Fuller, 117.
89 Ibid.
90 Ibid., 118.
91 Ibid., 119.
92 Ibid.
93 Ibid., 121.
94 Ibid., 124.
95 Alexander R. Galloway, *The Interface Effect* (Cambridge: Polity Press, 2012), 81.
96 Ibid., 82.
97 Ibid.
98 Ibid., 82–83.
99 Ibid., 84.
100 Ibid., 92–93.
101 I have written about these two artists in RACAR.
102 Galloway, *The Interface Effect*, 94.
103 Ibid.
104 Ibid., 99.

105 Bruno Latour, "On Actor-Network Theory: A Few Clarifications," *Soziale Welt* 47, no. 4 (1996): 369.

106 Ibid., 370.

107 Ibid.

108 Ibid.

109 Galloway, *The Interface Effect*, 80.

110 Jacques Rancière, "Are Some Things Unrepresentable?" *The Future of the Image*, trans. Gregory Elliott (New York: Verso, 2007), 109–38, 134.

111 Ibid., 135.

112 Laura U. Marks, "Infinity and Accident: Strategies of Enfoldment in Islamic Art and Computer Art," *Leonardo* 39, no. 1 (2006): 38.

113 Ibid.

114 Ibid., 40.

115 Ibid., 41.

116 Ibid., 38.

117 James Elkins, "Art History and Images That Are Not Art," in *Images: Critical and Primary Sources: Understanding Images*, ed. Sunil Manghani (London: Bloomsbury, 2013), 50.

118 Ibid., 49.

119 Why art history in relation to these images?—"Art history is centrally positioned in this emerging field because it possesses the most exact and developed language for the interpretation of pictures." See ibid., 51.

120 Ibid., 62.

121 Steinberg, *Other Criteria*, 84.

122 Ibid.

123 Sarah Milroy, "Francine Savard's Weather-Forecast Works Shine Bright," *Canadian Art*, 4 November 2013.

124 Steinberg, *Other Criteria*, 90.

125 Lagrange Paquet artist's website: http://www.eeelllppp.com/Histoires-d-interactions-v1-2014

126 Ibid.

127 Ibid.

128 Pasi Väliaho, *Biopolitical Screens: Image, Power, and the Neoliberal Brain* (Cambridge: The MIT Press, 2014), 33.

129 Margaret Iversen, "Index, Diagram, Graphic Trace," *Tate Papers* 18 (2012), available online: http://www.tate.org.uk/research/publications/tate-papers/18/index-diagram-graphic-trace (accessed May 5, 2017).

3

Celluloid Film as Digital Art: Translation, Information, and Intermediality in Cory Arcangel

Cory Arcangel is a multimedia artist best known for *Super Mario Clouds* (2002) [*Figure 6*]. Consisting of a projection of the entire *Super Mario Bros.* (1985) video game without the brothers, the mushrooms, the Koopa Troopas, or the obstacles—just the distinctively pixelated white clouds serenely floating on a perfectly blue backdrop from left to right—this work is not, as one would expect, purely digital. It straddles the digital and the material: the artist has cut into a Nintendo cartridge and physically hacked the circuits to modify the game. Arcangel qualifies the term "hack" by adhering to an older definition, wherein hackers felt their way haphazardly through code, rather than the newer understanding of hackers capable of erasing their traces.[1] Arcangel emphasizes his connection to the materiality of the object, not just its virtual content. For instance, in *Super Mario Clouds*, the Nintendo cartridge is rather bluntly cut into, the chip removed, and a new one refitted more or less in the place of the original. The modification also brings to light the materiality of the cartridge, the sight of which conjures nostalgia.[2] Nostalgia, in this case, doubles as a connection to a recent past and an essential stylistic device for the artist. The virtual space shown on the screen—deceptively simple clouds—taps into nostalgic feelings by mapping out a relatively new type of territory: the video game space. References to landscapes—such as the scientific observations resulting in Turner's famous cloud studies, or even Steiglitz's series of photographs of clouds, entitled *Equivalents*, signaling self-referentiality in a medium focusing on its limits—are inescapable, firmly planting *Super Mario Clouds* into an art historical continuum.[3] Arcangel's work opens up to many rich associations while simultaneously engaging pressing contemporary issues of intermediality, as well as those of place of technology in art and culture. It speaks about a technological history from the perspective of someone who has "cracked" the aesthetic "code" of the Internet and video games for a new generation whose constructed memories are of the virtual landscapes in which they dwelled for so much of their childhood.[4]

Super Mario Clouds is just part of Arcangel's extended media practice, which includes music, music videos, email-based art, drawings, kinetic sculptures, and even a novel based on other people's tweets. Recently, he has questioned the nature of the image in the digital age by emphasizing film in an attempt to exacerbate the rift between digital and celluloid images. While he sought to *bridge* the digital and the material with his Nintendo cartridge,[5] his new work seeks to *disassemble* the digital and analog. This chapter explores three such "film-hacks": *Untitled Translation Exercise* (2006), *Colors* (2006), and *Structural Film* (2007). *Untitled Translation Exercise* starts with Richard Linklater's *Dazed and Confused* (1993) and overdubs it with voices of workers at an Indian outsource firm, creating an uneasy disconnect between the portrayal of middle-class America and the perplexed tone of Indian voices. *Colors* (2006) appropriates Dennis Hopper's 1988 film of the same name, re-screening the gangland cop movie one horizontal line of color pixels, outstretched vertically down the screen, at a time. The film's figurative representations are replaced by an abstract digital image while the

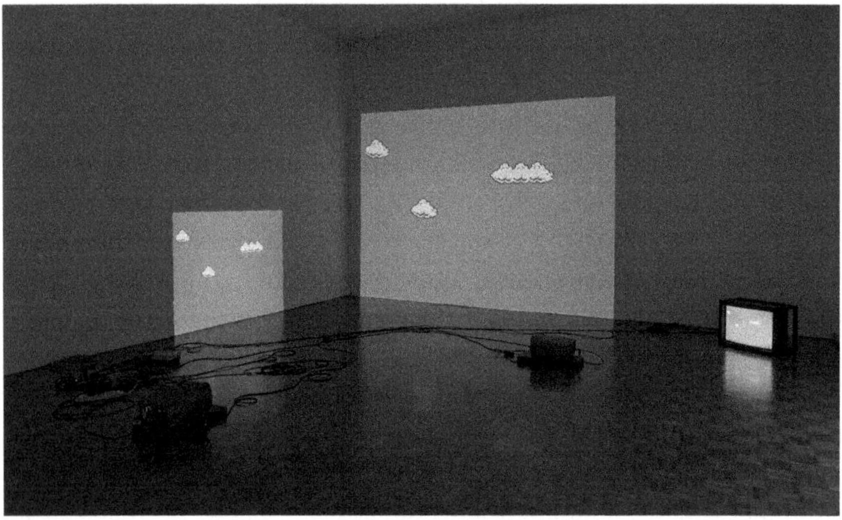

Figure 6 Cory Arcangel, *Super Mario Clouds*, 2002—. (Installation view, Synthetic, Whitney Museum of American Art, 2009.) Handmade hacked *Super Mario Brothers* cartridge and Nintendo NES video game system. Edition no. 2/5. Whitney Museum of American Art, New York; purchase with funds from the Painting and Sculpture Committee 2005.10. © Cory Arcangel. Courtesy of Cory Arcangel.

narrative soundtrack plays on. With *Structural Film* (2007), Arcangel projects a 16 mm film of a glitchy digital video filter to create a "fake" structural film while, in effect, offering a close study of the formalist tension between digital video and film aesthetics. These three works show the artist problematizing and intensifying the rift between the digital and the analog. I analyze Arcangel's film-hacks through the theories of translation, information, and intermediality. Rather than doing a comparative analysis of each medium's technical characteristics, I propose to map out the zone of indiscernibility between celluloid and digital images by suggesting a concept of the image based on varying degrees of formal and abstract arrangements rather than representation. In fact, these images play on paradoxical yet productive redundancies, scrambling codes in content and form to exploit failures in order to carve out an intermedial haptic zone. In the process, I bring attention to the new types of images developed by Arcangel, reinvigorating the aforementioned concepts so as to repurpose them for a visual aesthetic theory geared toward digital art.

Untitled translation exercise

The poetic, the unfathomable, and the mysterious can only be tapped into by a translator-poet, argues Walter Benjamin in "The Task of the Translator." Initially the preface to a translation of Baudelaire's writings, it is now a text of high importance to the craft of translation. There is an irony in this since Benjamin's text problematizes the very notion of the possibility of translation.[6] Indeed, it could almost be seen as an anti-translation tract. I use Benjamin's text to interpret Arcangel's *Untitled Translation Exercise*—itself, despite its title, a nontranslation [*Figure 7*]. Because of its odd status as translation, Arcangel's exercise serves as a metaphor for thinking about intermedia translation or intermediality. On the surface it seems to be a work of appropriation, but it is also a work of conceptualism with a global social component. Arcangel appropriates Linklater's *Dazed and Confused* to make modifications that seem benign but which, in practice, actually have far-reaching consequences.

Dazed and Confused follows several teenagers on the last day of school. Jocks chase cheerleaders, nerds get out of their comfort zone, virgins want to lose their defining trait. The film is a compendium of heterosexual, sexist clichés firmly associated with the myth—in the Barthean sense—of American high school with all of its underlying socio-economic white privilege. Of course, the high

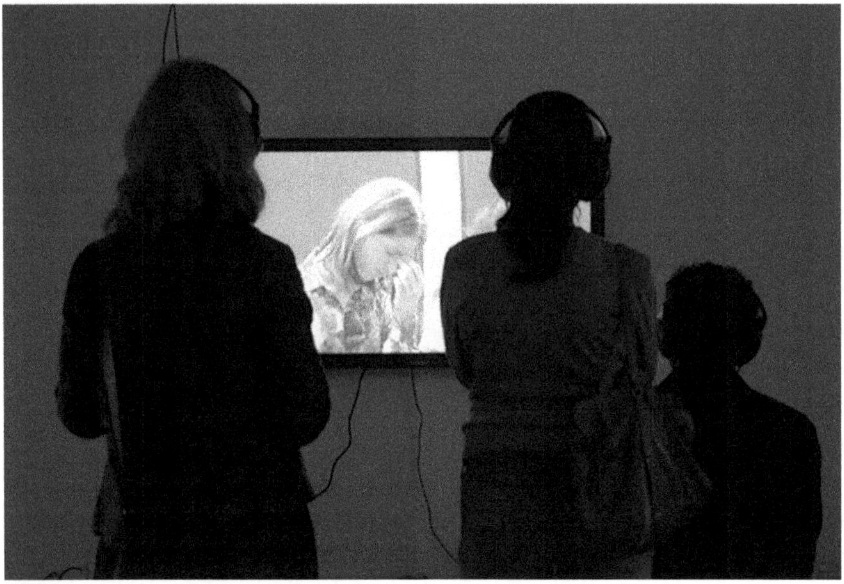

Figure 7 Cory Arcangel, *Untitled Translation Exercise*, 2005. © Cory Arcangel. Courtesy the Artist and Lisson Gallery.

school kids are all struggling with potentially life-changing issues: *Will I stand up to coach's authority, Will I sleep with the girl I like, Will I tell my father I want to be a dancer?* The problems here, even if they drive the narrative of the comedy, are nevertheless a social construction of the suburban American middle class. It is a feel-good film: everyone wins at the end by getting high, getting laid, and getting into a fist fight for the first time. And as the sun rises on a post-party landscape, these teenagers know the future is theirs for the taking: "Slow ride, take it easy," the words of Foghat, spread through the film's soundtrack.

To read Arcangel's video artwork of Linklater's movie through Benjamin's celebrated text on translation involves exposing the rift in Arcangel's translation through the notions of communication, embodiment, and data in Indian call centers. I then look at Massumi's concept of affect in his treatment of voice-over. The translation-which-isn't-one will push us toward an aporia, making J. Hillis Miller's "Paul de Man at Work," in which he provides an update on de Man's famous take on Walter Benjamin, indispensable. Finally, I will use Martin Heidegger to show the role of a bad translation in aesthetics and Gilles Deleuze to stratify it—code it, as it were—for further use.

Arcangel displays the film *Dazed and Confused* on a TV monitor in a gallery space but with one modification: the film has been "dubbed back into English via an outsource firm in Bangalore."[7] The film is "translated" into English after taking a detour through India. The trajectory insinuates that the original American language needs to be dubbed into English to be understood. The fact that this English has a decidedly Indian accent only serves to decode the confusion of the original. Arcangel commissioned workers at a phone answering company in India to read the script, and then dubbed the resulting recoding back onto Linklater's original film. The accents articulated by the voice-over "actors" (more like "voice-over laborers") sound conspicuous. The artist seemingly creates a disjunction between the English spoken by white American kids and the English spoken by Indian phone workers for a purely comic effect. The disruptive humor also emerges as a result of the fact that whereas the original English script was spoken by the actors themselves, the dubbed English, with call center agents' inability to see the context of the action, is tone deaf. This effect is manufactured, on some level, by having call center agents read a script in isolation without visual guidance from the film. But if humor was the goal of *Untitled Translation Exercise*, the work would hardly be worth analyzing. Instead, it is quite serious: as Arcangel warns his audience at Berkeley during a talk on the subject of his corpus, "in five minutes you will not be laughing."[8] What stands out in this "translation" is the confusion, the tentativeness, the miscomprehension in the voice-over track. What becomes apparent is the rift between the low-paid Indian voice-over crew and the superficial kids and their privileged lives on screen. It seems almost cruel, in the end, to juxtapose Indian voices with those of American teenagers. Not cruel to the Indian call center workers, but to the suburban teenagers, for whom you feel a deep embarrassment as they traipse around their unself-reflexive lives.

The incongruence between original and "translation" embodied in Arcangel's art piece seems to issue from modes of communication. The work seems to problematize the global circuitry of communication rather than offering communication as a solution to problems of globalization. Communication is significant because it is indicative of higher levels of social differentiation, as Harmut Winkler explains: "Communication, above all interrelates what has been separated by the division of labor, with regard to its content, its functions, but also its geographic implications."[9] Here, communication seems to suture the rift between languages on a global scale. Communication, according to Winkler, joins us together. And yet, Arcangel articulates a problematization of communication. Separations in labor division are not brought together but rather, and explicitly, make the geographic implications, to use Winkler's words, the "real" issue behind call centers. The problem of geography was an easy solve for communication: global space has been bridged since the first telegraphic message. Yet, the problem of geography emerges quite urgently in the context of affect and biocapital, which will be discussed below. One potential solution, displayed in Arcangel's *Untitled Translation Exercise*, is to forego meaning sought in communication in favor of intensity. Highlighting the theme of cultural rift further, Arcangel explains to his Berkeley audience that he had the call center employees record over a 100 minutes of *Dazed and Confused* script at the cost of $193.[10] An audible gasp from Arcangel's audience accentuates the discrepancy in pay scale in globalized labor conditions between Hollywood rates and those of call centers in India.

Kalindi Vora, in her text "Call Center Agents: Commodified Affect and the Biocapital of Care," explains the sociologically affective issues behind India's customer service industry, whose manpower is made up of English-speaking college graduates.[11] Vora explores the biopolitical deterritorialization that leads to an accumulation of surplus value capital outside India,[12] and in particular focuses on a type of globalized alienation afflicting this relatively new source of "affective labour."[13] The affective element denotes labor valued for such nontangibles as personality and emotional quality of the voice, and refers specifically here to the need for call center agents to develop a persona—a new name, fake accent, back story—to effectively do their job.[14] Essentially, the call center agents are actors reading a script, but somewhere the lines get crossed and communication becomes complicated—the callers and their interlocutors are processed as data: "The customer becomes a 'profile,' chosen by dialing software specific to the call center industry. The software that manages this digitized interaction chooses profiles based on algorithms that determine the highest match rating between the profile and the type of call being made, whether this is sales, collections, and so on."[15] On some level, from this data-based transmission, we are dealing with "specters," in Vora's words, composed of data forms constructed to produce affective commodities; these "ghosts" are derived from the manipulation of specific cultural knowledge with the aim of soothing agitated customers while rendering the caller, if only virtually, approachable.[16] Vora also mentions how the culture behind the call center is problematized on some level within their actual society: "They are also frowned upon because the youth culture around those who work in call centers—going out to clubs and bars after work in the early morning, dating, and otherwise emulating Western culture—has negative connotations outside of young urban social groups."[17] Here, they might actually have something in common with the disaffected teenagers of *Dazed and Confused*.[18] Arcangel's voice-over critiques the rift occurring in a global economy and the alienation brought about through communication technology. Much like a literary chiasmus, the viewer is exposed to the interplay of inverted social structures that reveal their sameness and that illustrate the rift within each youth's respective space because of the global economic and communication exchange.

But the link between these two groups of young people, the call center agents and the suburban filmic Americans, happens in a moment of transfer that relies on an exchange of an affective self through various technologies: "Affective labor and human biological materials also rely on specific technologies of extraction to be transferred to distant bodies. How value is carried and transmitted by affective commodities is an essential question for thinking about alienation as well."[19] How this process is enacted is really interesting from the perspective of translation, embodiment (or disembodiment), and information: "The transformation of the [call-centre] agent into her data form requires the suppression of her real form and yet results in the enhancement of the real form's life chances, because it gives her access to global flows of capital and labor demand."[20] Vora strikingly demonstrates how the opposite is true: "Her non-data form can contribute only by reproducing the life of her data form."[21] The interconnectedness between non-data and data forms comes from Vora's assessment of the slippage between biology and data. Information neutrality is what is at work here, where the content matters less than the expression of information: "The biological sciences are increasingly becoming information sciences, as what are perceived as the basic units of life, DNA, are translated into binary code and managed by computerized information systems. The production of biological life as information, or data."[22] Might this be the ultimate translation? Life as data? On some level these disembodied voices tentatively trying to approximate the images on the screen do seem like an emergence of a new type of life based on the virtuality afforded by communication networks.[23]

This notion of the voice-over must be further explored from the perspective of affect (already introduced by the nature of the labor of call center agents). Brian Massumi constructs his definition of affect on a psychological case study where the variation is the impact of a voice-over on the overall film's effect. The voice-over is yet another name we could give the translation of Arcangel's work, since it is an English film overdubbed into English and therefore comments on the film—a bad translation. In "The Autonomy of Affect," Massumi recounts the details of a case study where children are asked to watch three variations of a cartoon of a melting snowman: the first has no sound; the second has a factual voice-over; and the third has a voice-over accentuated with emotional words. The children liked the soundless version, disliked and had trouble remembering the factual voice-over version, but remembered the emotional one the best. Interestingly, the children reported that the saddest parts of the cartoons were the most pleasant to them. The children were wired so as to record their physical response: the factual cartoon (the least liked and least remembered) deepened their breathing and accelerated their heart rate the most, while their skin responded most positively to the wordless cartoon. Massumi concludes no connection can be drawn between an image's effect and its content, its qualities, and its intensity: "The level of intensity is characterized by a crossing of semantic wires: on it, sadness is pleasant. The level of intensity is organized according to a logic that does not admit the excluded middle. This is to say that it is not semantically or semiotically ordered … it vaguely but insistently connects what is normally indexed as separate."[24]

Why is this significant in the context of a work of overdubbing? Because the success of Arcangel's work resides precisely in this gap between content and affect. By doing this translation, not only from English to English but from movie to art, Arcangel is, following Massumi's logic, displaying the affect at play. But to understand the mechanics of affect, we have to look to the duality of intensity and qualification. Massumi explains how intensity and qualification are instantly embodied. "Intensity is embodied in purely autonomic reactions most directly manifested in the skin—at the surface of the body, at its interface with things."[25] Qualification—or the reaction to form or content—occurs at a deeper level (heartbeat, breathing) because, as Massumi explains, it depends on consciously being aware of narration and expectation of its continuity.[26] Arcangel enacts this idea in his translation piece: instead of being commentary on the unself-conscious privilege displayed by American whites in the Linklater film in relation to a larger global context, the overdubbing creates a disruption in our expectations of the narrative. Consequently, the narrative of the filmic medium is disrupted. A gap emerges: What to do with this zone between two media, two versions of English, and two forms of expression?

In his "Task of the Translator," Benjamin asks about miscomprehension: "Is a translation meant for readers who do not understand the original?"[27] An immediate question arises: Did Arcangel really believe we did not understand Linklater's film to such a degree that we required a translation?

For Benjamin, a translation is not simply a work presented again in a different language. Rather, according to his theory, the original work and its translation are fundamentally different things: "This would seem to explain adequately the fact that the translation and the original have very different standing in the realm of art."[28] Benjamin asks why one would go through the trouble of "saying the same thing" over again.[29] Does *Dazed and Confused* say something completely different in its modified, yet repetitive, form as *Untitled Translation Exercise*? A bad translation, Benjamin says, simply reiterates the same message of the original, so Arcangel's word-for-word reiteration of the original should be a source of the work's failure. And *Untitled Translation Exercise* seems to be such a bad translation that, even as it actually repeats the same words as its source, the original message is lost in translation. Thus, Arcangel, by literally reiterating the same text in the same language, translates something into a new form. Arcangel presses Linklater's film, with its cliché roles and nostalgia for the 1970s, toward a commentary on the current fearfully conservative view of white Americans feeling assailed by "barbarians" who can't speak "their" language. It is a reflection on the global economy that allows the Western world to wallow in its own consumerism while being blind to the work going into the production of goods. It asks about what services one can buy with money. It is about a fundamental global inequality. Arcangel asks through his translation exercise: Do you not understand the original? No? So let me translate it for you.

Reading Paul de Man on Benjamin's notion of translation is illuminating when applied to Arcangel's recent work. For Miller, de Man's sense of urgency is triggered by the message of the inhumanity of language he unearths below the strata of meaning in Benjamin's translation theory. (One can see already that the threading through of theorists on the subject of translation puts into motion the idea of translation as a methodology enacting an interpretive stratification.) Can inhumanity be also uncovered in Arcangel's light-hearted art pranks?[30] Arcangel drives at a gap dividing old and new media by wedging a distance in time (from the 1970s to now), space (between India and the United States), and levels of expression (movies and art). But does the gap in Arcangel's work open to Miller's bleak reading of translation? Miller explains how the failure of translation is the result of what is meant and how it is meant: "Far from being a matter of human intention, this incompatibility is a consequence of the inhumanity of language."[31] He suggests that we should be suspicious of language in a time when we need it the most.

> Why is that important now, when we have other things to worry about, from the melting of the Arctic icecap to global financial meltdown to the meltdown of the humanities? My shorthand answer is that de Man was prophetically aware of the way assumptions about "the human" and about related concepts such as pan-organicism can get us in big trouble. Paradigmatic within aesthetic ideology is the assumption that language is human and within human control, whereas language, as de Man patiently showed by way of what Benjamin is really saying, is an inhuman machine. Language is a machine that, performatively, *verspricht* (*sich*), falsely promises and contradicts itself at the same time.[32]

Miller's inhuman machine is an indictment of the power of communication. De Man, following Benjamin, sketches out the difference between translation and poetry. Whereas poetry does not transcribe, imitate, or paraphrase, a translation is rooted within language: "Translation is a relation from language to language, not a relation to an extralinguistic meaning that could be copied, paraphrased, or imitated."[33] Arcangel manages to say something like the poet, paradoxically by showing the division in the relationship of language to language, or in this case English to English. The "extralinguistic," or affective, meaning comes from the disjunction or what Massumi explained as the curbing of expectations. "You can translate only an original."[34] By not translating it really, Arcangel shows the clichéd nature of the paradoxically original movie, or already slippery nature of seemingly hermetic boundaries between media. And translating also brings other problems for the original. De Man draws similarities between translation and criticism or theory insofar as they are ironic gestures. The purpose behind irony in the gesture of critiquing/theorizing is the same as translating, that is, to destabilize the original. The importance of the original is underscored *because* it is translated. But then, the original is foregone in the importance-bestowing translation or critique. Ironically, the theory of the object matters more than the object.[35] The original is not definitive, says de Man, since it needs to be translated. Translation ironically undoes its importance by playing off this very need. The irony written about by de Man is clearly visible in Arcangel's work. It is paramount to the success of the piece. Here, though, Arcangel translates ironically by not really translating; in doing so, he undoes the perceived stability of the original. By extension, he destabilizes the medium of film, whose images are transmediated, wrenched from their material origin.

To reiterate this point: according to de Man, critical philosophy, literary theory, and history—as activities—"do not resemble that from which they derive."[36] Instead, "they disarticulate, they undo the original, they reveal that the original way always already disarticulated."[37] Isn't this what Arcangel *Untitled Translation Exercise* does exactly? Besides, wouldn't Arcangel's art as critical translation fall within the grouping of "activities"? Arthur Danto, after all, once defined art as art theory.[38] In fact, Danto carved out this definition of art as art theory from a contemplation of the nature of the original and its imitation. De Man goes further: "They kill the original, by discovering that the original was already dead."[39] This tautology is the point of *Untitled Translation Exercise*. It reveals the retro corpse of middle-class America that was already its own nostalgic specter in the 1990s when the film was shot.

But translation can be even bleaker for de Man: "Translation, to the extent that it disarticulates the original, to the extent that it is pure language and is only concerned with language, gets drawn into what he calls the bottomless depth, something essentially destructive, which is in language itself."[40] Who knew that translating an English film to English would be ruinous to the viewing experience. Does the voice from India calling us on the phone sound like a signal from a bottomless depth? In addition to the alienation due to communication in global markets, consider also the alienation de Man digs up from Benjamin's translation theory. The point was already made with the alienation that occurs in the global markets through communication, but here de Man also shows how this alienation is present at different levels of language: "What the translation reveals is that this alienation is at its strongest in our relation to our own original language, that the original language within which we are engaged is disarticulated in a way which imposes upon us a particular alienation, a particular suffering."[41] Maybe in the film we see the economic seeds eventually growing into a global crisis manifest in call center alienation. And this is the ultimate reveal of Arcangel's disruptive nontranslation. Like a Heidegger photograph of the masks of the dead, Linklater's film is a dead photograph or a dead era that never existed nonproblematically. *Untitled Translation Project* illustrates that, as Benjamin says, "a literal rendering of the syntax completely demolishes the theory of reproduction of meaning and is a direct threat to comprehensibility."[42] Arcangel's work clearly illustrates how a translation translates language and not text: he is using the same text yet making it say something else. But how do we map out this rift in language, between translation, between media?

Can translation across language or media platforms open a space, a critically considered boundary line? Can translation—to be more precise, self-reflexively, as in Arcangel's work—offer clues as to what happens to an image translated between media? The concept of rift (*Riss*) is essential to Heidegger's theory of language and aesthetics. For Heidegger, the issue of translation is one of artistic form. How we understand art is in part based on the rift in translation between ancient Greek and Latin. Heidegger, in his *Origin of the Work of Art*, names three different ways of approaching art before settling on the equipmentality: structure, aesthetics, and hylemorphism. Structure is defined by a thing and its characteristics and is based on language—in particular sentence structure and its relationship to the thing it claims to describe; aesthetics is based on sensation; and hylemorphism on form and matter. Each of these ways of approaching art is discarded by Heidegger for various inadequacies. The first is rejected on the basis of a bad translation from Greek to Latin. This rift is, according to Heidegger, responsible for "the rootlessness of western thought."[43] He perceives the problem in the inevitability of structuring a theory of art on language. He writes, "Beneath the seemingly literal and thus faithful translation there is concealed, rather, a *trans*lation of Greek experience into a different way of thinking."[44] There is a detachment between words and experience, or, as it were, language-based alienation. Experience is not transmitted in the translation. And yet, it is experience, albeit in a very different form, that makes itself apparent in Arcangel's artistic repetition. It is this unfamiliarity in the differing experience that makes this translation, or nontranslation, successful in Benjamin's eyes and instrumental for Heidegger, who claims this unfamiliarity, this rift in experience, as the impetus to "think and to wonder." Benjamin also wondered about the form, or the packaging, of the element of translation: "The translation must be one with the original in the form of the interlinear version, in which literalness and freedom are united. For to some degree all great texts contain their potential translation between the lines."[45] Benjamin is reading between the lines, probing the intra-linear; the unfathomable aspect of a successful translation is captured in the zone between the lines, devoid of content, empty of intention.

This space of translation is picked up by Deleuze and Guattari in *A Thousand Plateaus*. It is a zone, a space in-between the lines. But the lines are material striations. Here, translation becomes a way of thinking about the movement between two systems. The strata, striated, linear horizontal platforms, demonstrate a way of codifying information that can be seen as skipping from one system according to incorporeal functions: "The temporal linearity of language expressions relates not only to a succession but to a formal synthesis of succession in which time constitutes a process of linear overcoding and engenders a phenomenon unknown on the other strata: translation, translatability."[46] For Deleuze and Guattari, translation is a question of moving from one system of representation to another. The notion of translation is based on function, and this very spacing between two different systems takes on a form. And so, we are here near to what Benjamin had in mind when he was looking for translatability between the lines and what Heidegger augmented in terms of a rift leading toward aesthetics. Deleuze and Guattari write: "This property of overcoding or superlinearity explains why, in language, not only is expression independent of function, but form of expression is independent of substance: translation is possible because the same form can pass from one substance to another."[47] So how do we move away from this notion of translation based on content to translation based on function at the level of form? How do we continue to search for a zone between media that is, by all accounts, indiscernible? A good place to continue is in the notion of streamlining of information, or in more concrete terms, the abstraction of figurative images into geometric abstraction in Arcangel's *Colors*. The shift is between metaphorical lines illustrating codification and the organizational space of two language systems to actual, and colorful, albeit not static, lines.

Colors

Colors (2006) is Arcangel's appropriation of the movie *Colors* (1988) played one horizontal line of color at a time [*Figure 8*]. The identical title functions here like the English-to-English nontranslation: ironic critique of form and content of the original. More precisely, the artist transforms filmed images in video form based on an algorithm he has provided for the public on his website. The source is the movie *Colors* but it is unrecognizably transformed into an *objet d'art*. The original movie is directed by Dennis Hopper. It is a police drama revolving around two cops played by Sean Penn and Robert Duvall assigned to an Los Angeles gang unit and their differing styles in dealing with two gangs: the Bloods and the Crips.

Duvall is experienced, steady, and cool; Penn is hot-headed, violent, and impatient. The colors are not only referring to the black and white of the racial divide, the blue of the Crips, and the red of the Bloods, but also the blue of the police uniforms and the red stains caused by gunfire. Multicolored murals, graffiti, and artworks as backdrops intensify the scenes. The use of primary colors—yellow

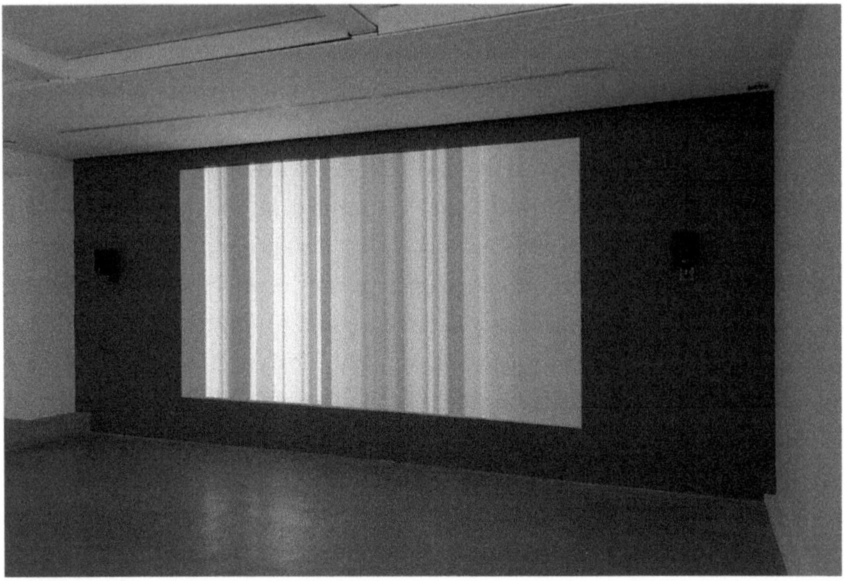

Figure 8 Cory Arcangel, *Colors*, 2006. © Cory Arcangel. Courtesy of the artist and Lisson Gallery.

"Pac Man" car, red blood, blue jeans—makes the scenes pop. But maybe, apart from the title of the work, what is significant in the film is not the colors, both figural and symbolic, but rather the outdated 1980s street slang and Ice-T's synth hip-hop. The image is reduced to vertical strips of color, leaving the soundtrack as the only recognizable aspect of the film. The work by Arcangel suggests a continuity with the narrative elements of the film: "Hopper's narrative structure also suggests the stubborn resilience of codes of color, stereotypes, tropes when the true reality is much more complex, and yet these easy patterns are repeated over and over again."[48] The art work reduces the stereotypical patterns of the narrative film into coded color patterns—the codes of the street become codes of programming. In effect, rather bluntly, the artist makes an analogy between the conventions of social coding and the informational equalization of digital coding.

In this section, I first compare *Colors* to Guido Molinari's paintings from the 1960s and 1970s. Molinari reduced the color field to vertical stripes by researching the link between vision and understanding in structuralist theory. Structuralism, despite being an antiquated apparatus, can still offer some clues about the relationship between abstraction and visual information. Even though the comparison between two artists who happen to use stripes seems to be made on the surface level, it opens a discussion into the nature of information and the visual medium. First, this discussion is prompted by Molinari's use of structuralist theories to inform his art practice. Second, I examine the nature of abstract information and its relation to art.

Colors was Hopper's second directorial effort, the first one being *Easy Rider* (1969). It is noteworthy that Hopper was also an artist who transferred images from one medium to another, specifically, photography into painting. He was also a practicing abstract artist. But Arcangel's radically modified version of Hopper's movie is far from being an homage to the actor/director's artistic career. Rather, the look of Arcangel's piece is more of a visual analog of Montreal minimalist stripe painter Molinari. Molinari started doing paintings consisting exclusively of stripes in 1956.[49] A Molinari could thus be facetiously considered as a photorealist still for Arcangel's *Colors*, which, in turn, could be construed as a live-action rendition of Molinari's paintings. For Molinari, the stripes were a solution to a vertical/horizontal duality as the source of the problem of mass, gravity, and trajectory within the internal structure of the picture plane. Molinari believed that vertical/horizontal duality reinforced the idea of closed space. This vertical/horizontal opposition is at the base of a Euclidean conception of space with an already built-in perspective, which, in turn, implies figurative representation.[50] The stripes solved that ingrained problem.

For Molinari, the issue revolves around an expressionist element he discovered in Mondrian's opposition between color and smaller-scale rectangular forms. According to Molinari, this opposition expresses the relationship between individual elements and a whole, providing diegetic fundamentals. Molinari set out to eliminate this conflict between object and space and, at the same time, the tension between multidimensional objects in relation to each other. In doing so, he avoids a horizontal line, the minimal element suggesting a landscape. Hence, he settles on stripes: their similar width eliminates the expressionism born out of a difference in proportions. They are then dependent on the qualitative function of the variation in color from one stripe to another positioned in the sequence that fills the canvas. Their vertical repetition creates a rhythm that, according to the artist, creates an illusion of space.[51] This is not an analogical space, however, which is something Molinari wanted to eliminate, but a haptic space, a term we will get to. The illusion is uncanny: one only has to stand in front of a striped painting, for example *Bi-sériel orange vert* (1967), to see a slight trembling movement opening a perceptible space between stripes: it is as if the canvas were in motion. It is a space resulting from a mix of digital and analog—haptic, intense—since the artist tried to supplant the medium's reliance on analog representation.

Molinari rejects the figure/ground duality in painting based on his studies of contemporary philosophy of science.[52] Despite their outward similarities, video stripes and paint strips are essentially quite different. To shuttle from one medium to the other or one milieu to another, we need code. Abstract painters, Deleuze explains, do not simply apply an external code to painting, they rather "elaborate an intrinsically pictorial code." Deleuze admits that "it is a paradoxical code." Coding painting is already playing in the interstice of the analog and digital: "It is the digital expression of the analogical as such … It is as if the diagram were directed toward itself, rather than being used or treated." A code is necessary to a meaningless abstraction: "It no longer goes beyond itself as a code, but grounds itself in a scrambling."[53] The result, like an old television set losing the signal, is a visual static. Digital code will bring data to a homogeneous whole and then binarize it on a separate plane: "Abstract painting obviously proceeds by code and program, implying operations of homogenization and binarization that are constitutive of a digital code."[54] Translation comes into play. Analog is close to intensity, adding and subtracting—modifying as it were, that which is the diagram. Whereas for the painter the codification is something achieved on the diagrammatic level, somewhere between gesture and brush stroke, Arcangel's more literal code is shared online. Is the process that Molinari went through akin to figuring out an algorithm, a formula? The painter's process was to rationally vet an artist for inspiration, consult visual researchers, and calibrate theories to arrive at a signature style. The stripes were meant to make familiar organization of the picture plane unrecognizable in a clear program of scrambling.

Molinari, as explained by Robert Welsh, went through a period of carefully choosing the alignment his style of abstraction was going to take: Pollock or Mondrian. To inform his decision, he also looked into the writings of Ernst Mach, known for his Mach Bands, where the edges of gray bands in a series of gray bands seem to have a slightly different shading than the rest of the bands; and Franz Schumann, who is known for his optical illusion studies based on grid lines of differing groupings. To refine his choice, Molinari also sifted through the structuralist theories of Charles Biederman, an abstract artist and theorist; philosopher Alfred Korzybski; and psychologist Jean Piaget. Just like the optical researchers, the three structuralists—the art theorist, the philosopher, and the psychologist—could be said to be interested in lines.

Biederman, for example, writes about the importance of the line in the emergence of art in his *Art as Evolution of Visual Knowledge*, published in 1948. He writes, "The line offers the simplest and most direct possibility for man's discovery of a medium and invention of a method of art."[55] As part of this evolutionary theory of art, Biederman sees the line and its manifestations as a sequential progression into the development of representational meaning and symbols: "Thus the linear artist began by putting lines together, producing groups or arrangements of them which were at first the result of invention. In the process he must have come to attribute to them symbolic meaning, even before he discovered that they could be converted to *representational* functions and meaning."[56] His treatise on the evolution of art eventually tackles the problems of art and sciences and suggests that Mondrian is in denial of the structure of nature. This would be of interest to a painter like Molinari, who was trying to get rid of traces of the natural reference in his paintings. But for Biederman, the work of the artist in relation to science is one of relation with reality. Science revealed nature's reality by lifting the veil of optical appearance.[57] By going beyond appearances, Biederman believes that we can get to an equalized, or unified, field of information. The artist has to then focus on the concrete level of things, like tightening the focus on a microscope: "Does not the change from the 'concrete' to the atomic reveal nature as a *creative reality transformations* from one level to another? Do not art and science disclose nature as a many-faceted reality each a part of nature's entirety, one grand creative progress?"[58]

For Korzybski, as for Biederman, the line, reality, and art are also interrelated. Lines, following Korzybski's notions based on "matter, space, time, mathematics," are connected to "real space" in relation to the "illusionist space of the painting."[59] Korzybski's working principle of abstraction hierarchizes perception. The appeal to an abstract artist is immediately recognizable in the way Korzybski considers the nervous system as "an abstracting mechanism."[60] The theory states, for example, that an object observed from the perspective of its molecular structure is appreciated at a lower level of abstraction than when it is encountered in its daily setting.[61] The high level of abstraction of the daily encounter with an object can lead to semantic confusion or misperception of reality. Furthermore, it is the consciousness one has of the levels of abstraction that helps dissipate semantic confusion.[62] Thinking in terms of visual illusions can help disentangle daily illusions: we see a solid disk instead of individual blades when a fan is turning.[63]

If we were to apply these principles to Arcangel, the artist could come out as an experimenter of the image: reiterating structuralist theories in a moving medium. This is especially true of an artist who is making us aware of the levels of abstraction in semantic illusions. An illusion is underscored paradoxically by turning stereotypical color codes into meaningless stripes, like multiple blades of the fan scrambled into a seemingly solid disk. In exacerbating daily abstraction through further abstraction, we see the need for clarity. For Korzybski, we can solve the spatial reading of levels of illusions by making the mechanism of the conceptual levels visible through configuring them according to a "diagram/blueprint."[64]

Korzybski's spatial thinking is echoed in Piaget's view of reality. Molinari follows Piaget in denying the linear connection between perception and operative intelligence and stating that "abstract thought images are derived directly from perception."[65] Molinari's work can be motivated by "a basic awareness of intuition of space" and seen as an "action performed on properties of objects rather than a mere reading of such properties," which here echoes Massumi's indictment of structure through the event.[66] The painted stripes, following this, would be more about what they do than what they mean. The actions performed by these objects' properties in turn produce "operational schemata which are then formalized."[67] This brings us back to Korzybski and the diagram/blueprint of configuration of a mechanism, in this case, of vision. But just as translation had its somber side, so does structure: "For structure is the place where nothing ever happens, that explanatory heaven in which all eventual permutations are prefigured in a self-consistent set of invariant generative rules."[68] As Massumi explains, structure is too rigid to accommodate intensity.

This process of producing work serves to see the liminal space between quality and intensity as explained earlier by Massumi: to "stimulate a pre-operational, pre-Euclidian … experience of space according to motor-sensory rather than purely perceptional processes of analysis."[69] This affect-based knowledge is what Arcangel's work brings to the surface, albeit with some degree of irony (since we are still hearing the movie play in the background, behind the vertical veil as it were). And so Arcangel's work, following the premise of Molinari's research that led to the conceptualization of painting as stripes, is an exploration of visual information and the limits of aesthetic production.

But for artists, this reliance on structure comes with a dark side. Eve Meltzer explains, in *Systems We Have Loved*, how the "structuralist imaginary" helps artists cope in the information-prison-house; instead of escaping the grid, they can recombine the code: "Just as 'information' at this movement drew much from structuralism's imaginary, structuralism, in turn, leaned heavily on notions of the informational."[70] The example Meltzer provides is Frederic Jameson's *Prison-House of Language*, which describes the notion of binary opposition much used in structuralist theory. In it, she explains, the subject experiences communication as a machine. Communication for Jameson is a "'technique for simulating perception,' necessitated 'when faced with a mass of apparently homogeneous data to which the mind and the eyes are numb.'"[71] The binary opposition principle is used to decipher or decode vast amounts of raw data that force us to perceive "difference and identity in a wholly new language the very sounds of which we cannot yet distinguish from each other."[72] This notion of the subjects being forced to attune themselves to the frequencies emitted from this undistinguishable soundscape is also addressed by Meltzer, this time through an even more somber Piaget, according to whom structure "inflicts a kind of death" on the individual subject in favor of an epistemic subject or "cognitive nucleus which is common to all subjects at the same level."[73] This death of the individual subject, coupled with the subject–machine, brings us close to this idea of the leveling effect of information theory. It should be noted that Hopper's individual subjects die in a shoot-out blood bath.

For Meltzer, the coded epistemic subject leads to the grid: "Binaries, … data, devices, modes of decoding, and deciphering … all evoke in striking similarity the image of the grid. But not just any grid. This is a closed system of synchronically occurring oppositional terms."[74] Or we might say, not just any grid, but one that is depicted by Molinari, structured by codes of language, or the one established through coding by Arcangel.

Seen from the perspective of information theory, a narrative film of cops and robbers is pretty much the same as a series of stripes. Carolyn L. Kane calls information the "new common denominator" and explains the process behind this reduction of content into abstract forms through how cybernetics views information: "All communication and cultural processes could be analyzed, viewed, and understood in terms of data and pattern formation. All humans, animals, and machines were herein treated 'equally': as media technologies capable of analyzing, sorting, transmitting, and processing information."[75] What is interesting here is how this view of equalization recalls the call center workers who have been processed like data. The ultimate translation of humans into data is, at least symbolically, shown in *Colors*. The experience of human figures of Hopper's film as a series of digital bar codes—ever changing in order to adequately respond to the complex reconfigurations occurring with each scene—illustrates the data translation of call center workers from non-data selves (figures in film) to data selves (coded stripes). This line of thinking applies to *Colors* because the work speculates on the notion of information and its essence: When does the film cease to be a film? When is abstract art devoid of content? We can't see the film, but we can hear the abstract work. There is synesthetic confusion that highlights the difference, the rift, between media and, paradoxically, the lack thereof in a new digital landscape—where information constantly flows through different platforms. Put simply, *Colors* is essentially an abstraction of human narrative into codified, non-narrative geometric art. It is a symbol of the same kind of stereotypical abstraction the film makes between whites and blacks or between blue and red.

But to understand this we must turn our attention not to content or form, but to the nature of information. Information theory, as Kane explains through Claude Shannon's 1949 Bell Laboratories experiments, "quantizes data in order to make communication processes more efficient."[76] Is a striped form of representation a more efficient way of seeing the cop movie? Does this help us read between the lines in a more literal way than Benjamin suggested? This efficiency, Kane continues, "is accomplished by separating redundancy, repetition, and as much noise as possible from an encoded signal so that it may travel swiftly and efficiently through numerous interchangeable channels."[77] What happens with this streamlining of redundancies into more efficient frequencies is a devaluation of meaning: "Because information theory quantizes data and information flows, concepts like 'meaning' or 'purpose,' normally given great weight in cultural and historical analysis, are abstracted into statistically calculable 'units of measure.' N. Katherine Hayles has argued that information herein 'lost its body.'"[78] This loss of the body through abstract information in the translation of meaning to an undifferentiated streamlining is exposed by Arcangel's *Colors*: the figures are erased from the screen, elongated into strict but random patterns by an aleatory set of principles (each pixel on the horizontal line descending through the grid of the screen vertically, systematically). But Kane warns, "information 'must not be confused with meaning … in fact, two messages, one of which is heavily loaded with meaning and the other of which is pure nonsense, can be exactly equivalent.' In information theory, the system only knows what it parses, processes, and orders as information, all else is 'noise.'"[79] This processing function of the screen is captured by Väliaho, who looks at the double meaning of "screen" that refers to: "both a material instantiation of the images it displays and a sort of sieve that selects what kinds of images will emerge in our cognitive reality."[80] The lines in Arcangel's piece actually reinforce our perception of the screen-object as an information delivery system.

Kane also reminds us, however, that "the line between information and noise is a precarious one."[81] For example, Massumi, discussing the idea of intensity and affect, describes it as a noise: "Intensity is qualifiable as an emotional state, and that state is static—temporal and narrative noise." If we were to translate intensity into the logic of emotion, it would not register: only static noise would be perceived. Intensity is something not easy to put into words. It lays dormant to cause confusion: "It is a state of suspense, potentially of disruption." Intensity is out of time; it is not something that fits a narrative; it doesn't transfer into a story: "It's like a temporal sink, a hole in time, as we conceive of it and narrativize it." Intensity is not passive either: as you get closer, you see the vibration of a teeming movement: "It is filled with motion, vibratory motion, resonation."[82] Vibration illustrates the notion of intensity as it can be perceived in art. It is not aimed at something in particular; it has no value, no meaning, no practical existence—except, Massumi adds, on the screen.[83] This vibration is perceived in Molinari's work between the lines. We could say that hints of intensity in the viewer's reception were already witnessed by Welsh when he suggested that, in the case of Molinari, "the entire painting is transformed into an event of visual and temporal energy vibrations through each viewer's system of perception."[84] This event breaks us out of the structure of the painting and involves a vibration that, for Deleuze, is the destiny of a painting: "This, finally, will be painting's great moment, its continuous movement, its vibration of vibrations."[85] These vibrations, central to the notion of sensation, carry the color that we experience in a Molinari or an Arcangel. The vibration itself is the interface of our reception of color. It is convenient that such reception in Deleuze is both digital and analog: haptic. The haptic is not precisely or directly the relation between touching and seeing. It is the all-encompassing experience of seeing and being aware of seeing.

> [The haptic is] a tactile relation with the optical or visual that neither subordinates touch to sight (as in digital vision, where we can choose or touch what we see based on predetermined alternatives, as on a computer or video game), nor subordinates vision to touch (as in purely manual referents which scramble or dismantle the visual); rather, a relation that shifts from manual (or analogical) referents within an optical (or digital), codified space.[86]

The haptic is the best way to approach Arcangel's digital image. It is like a veil, beyond which we can hear the film playing. The analog medium has been digitized so that what we are watching is manifest intensity.

Structural film

Colors paradoxically made the notion of noise prominent. It is a noise intensifying into a vibration that finally reveals the nature of the digital image from the perspective of all-equalizing information. But the idea of the in-betweenness of vibratory oscillation mediating two states becomes manifest on the practical level of the medium. *Colors* is not only about intensity emerging from the veiling lines of the narrative-obfuscating striped-screen; the piece also stakes out the borderline drawn between media. As Villaseñor explains, this negotiation between media is seen on the very surface of the materiality of the image:

> Arcangel's *Colors* (2005) … seemingly takes a Hollywood movie and transforms it into pulsing lines of pure, abstract color, but with one notable difference: the lines are generated from the digital material of the films themselves. One could argue that a radical transformation has already occurred when something shot and released on celluloid film has been digitized, but this too, is part of the reflective experience of Arcangel's *Colors*.[87]

Abstraction leads us to question the notion of intermediality. And the intermediality broached by *Colors* is that of the translation between celluloid and digital formats. I examine this here with another example of seeming intermediality in Arcangel's art.

Structural Film (2007) pushes the media boundary issue to the fore [*Figure 9*]. Here, Arcangel has in mind the films of the structural film movement with the likes of Michael Snow's *Wavelength* (1967), which was filmed on random film stock coming from canisters at various levels of material integrity.[88] Arcangel explains the nature of *Structural Film*: "Somewhere along the way on this one, a file got corrupted in one of the transfers, and some bits of colored stuff showed up, anyway, I kept it in the film, but those weren't actually part of the plan."[89] Arcangel is working on a very thin edge between digital and analog. The content of the film, white screen glitches, is more in keeping with Nam June Paik. While the subject matter is, as the title suggests, modernist aesthetics, these aesthetics

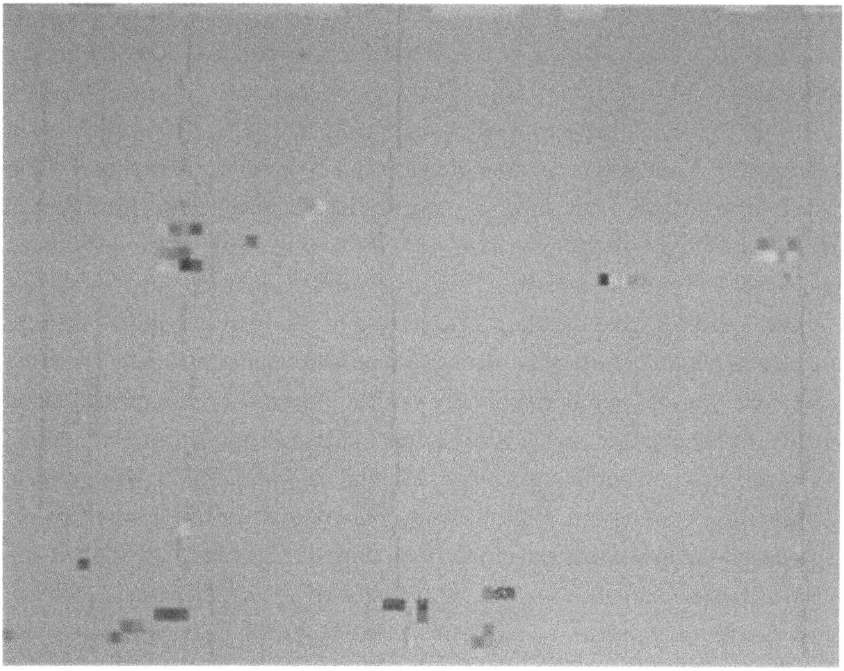

Figure 9 Cory Arcangel, *Structural Film* (still), 2007. 16 mm film. 6:15 minutes. © Cory Arcangel. Courtesy of Cory Arcangel.

are achieved through the transposition of medium-specific glitches onto a whole other medium where material errors should not have occurred and that, in effect, makes them aesthetically valuable.

Paik's *Zen for Film* asserts its celluloid materiality as it accumulates dust and shows the passage of time through the glitches that appear on the film stock[90]; the work shows signs of deterioration, such as smudges and particles.[91] A similar display of the ravages of time seems to be missing from Arcangel's work. And if the content of the projection—the glitch digital filter—appears immutable, the film stock, which projects the digital content, will eventually deteriorate. This negotiation between the celluloid and the digital has been recently waged against the digitization of *Zen for Film*. The analog film was displayed for countless hours and the traces it accumulated were eventually brought to a halt when the content of the analog film was digitized.[92] With its digitization, part of the artwork's conceptual apparatus was lost. But its afterlife as a digital work was not without glitches: "The digital display in turn reveals traces other than just scratches, dust, and chance events, in other words, digital forms of decay."[93] Perhaps *Structural Film* will become the "fossilized filmic artifact" if it follows *Zen for Film* down the intermedia translation road.[94]

The glitch issue in-between media can be compared to the notion of failure in translation.[95] Arcangel's translation of images leads to visible corruption even in a medium that should not show any traces of fallible materiality. This failure of translation brings us back to de Man, who invoked the issue of failure between an original text and its translation.[96] Arcangel's acts of appropriation of Linklater, Hopper, and Paik have the effect of elevating to the level of high art the first two objects while continuing the dialogue about intermedial materiality with the latter. The "gauche" appropriations are critical/theoretical readings of original visual texts based either on failure of content (bad translation), form (scrambling of image), or both (framing glitches in a failing medium). Through this critical or theoretical reading of a filmic artifact, "the original work is not imitated or reproduced but is to some extent put in motion, de-canonized, questioned in a way which undoes its claims to canonical authority."[97] What the actual canon can bring to art is far from our concern here. Where Linklater's or Hopper's films lose their "canonical authority" is where their ingrained worldview is deterritorialized by the digital treatment—the material translation of film to digital artwork mirrors a translation between ideological materiality and informational materiality. We no longer see them as they were intended to be seen. The new digital layer Arcangel adds to the film serves only, as de Man says, "to understand the original from the perspective of the translation."[98] The source films of *Untitled Translation Exercise* and *Colors* are fossilized by Arcangel as artifacts of the original, their content undecipherable runes for present audiences.

But where are we in the exploration of the glitch or failure that manifests itself in the original digital medium and how it is presented in 16 mm projection? If Paik was questioning the essence of the celluloid medium, Arcangel again frames it critically through an intermedial playfulness. Explaining how glitches are the embodiment of a software aesthetics, Peter Krapp, in *Noise Channels*, writes about the presumed infallibility of digital media and the alienation of the user: "As our digital culture oscillates between the sovereign omnipotence of computing systems and the despairing agency panic of the user, glitches become aestheticized, recuperating mistakes and accidents under the conditions of signal processing."[99] As the chasm equalizes, can this glitch aesthetic be a balm for posthuman tendencies? Is the glitch in Arcangel's work an inoculation against the spread of the posthuman denounced by Miller as the inhuman of conventional language?

> In the age of cybernetics, it can seem as if human fallibility is what keeps systems from achieving their full potential—from systematic closure. Yet rather than our becoming abstractly "posthuman" in information society, one might instead argue that people, citizens, and individuals in fact become realized for each other and for themselves in unprecedented ways through networks of computer-mediated communication.[100]

The glitch is a manifestation of the human. *Structural Film*, far from presenting an empty screen with glitches, in fact projects the specter of the human haunting the machine. In a final analogy, Krapp opens up a question of intermediality: "One might conclude, however provisionally, that gaming glitches are part of the art form in the same way that brushstrokes are part of painting."[101] The discussion of painting within the context of intermediality[102] brings us back to our theoretical painting/digital cross-pollination between *Colors* and Molinari, which can be carried over here as a general example.

What is this notion of intermediality? Why not simply speak of two different media and the mutual referentiality that occurs in interdisciplinary arts? It is because intermediality can function as critical translation in the de Manean sense: "Intermediality is often viewed as having the ultimate goal of 'figurating the infigurable,' the incommensurable."[103] By shifting from one medium to the next, Arcangel captures a snag that illustrates the incommensurability between the media. What it reveals is that what constitutes intermediality is not only the smooth relation between media but also the interference resonating between them.[104]

Intermediality can be understood as a critical category, one that, in historicity and constructedness, can hold a meaning.[105] If, according to Irina O. Rajewsky, three intermediation groupings can be isolated (medial transposition—i.e., novel adapted to film; media combination—i.e., opera or computer art installation; and intermedial references—i.e., reference of painting to film or painting to photography), one of them applies to *Structural Film*, which is the notion of "film qua medium" through references. Photorealist painting is not the result of a plurality of media coming together to create meaning. Photorealism, she writes, "it is not two or more different forms of medial articulation that are present in their own specific materiality. Instead, what we are dealing with is nothing other than painting—but a kind of painting which inevitably evokes in the viewer the impression of a photographic quality."[106] The second medium is evoked indirectly, whereas the primary medium is creating the illusion. To put it more clearly, Rajewsky explains further: "It is not photography which manifests itself materially; rather painting's *own* instruments and means are applied and shaped in such a way that experiences, or 'frames', are evoked in the observer that are medially bound to photography, leading to an illusion, an 'as if', of a photographic quality." Rajewsky shows how only one medium displays the limits of its "materiality and mediality."[107] Her theory can be used to identify the oscillation between celluloid film and digital art in *Structural Film*. What Arcangel is creating is a film projection of a digital glitch that itself came from a digital medium referring to film (and an "aged" film at that, as the filter is called). The level of self-reference could lead a skeptical viewer to interpret this process ironically, as an ironic "art film." And yet, the art historical sources are here quite overt (Paik). The resulting referentiality is simply part of the—unironical—mediated condition of the image. The very corruption of the file indicates a materiality beyond the actual. A process of interference between media.

The film projects clusters of colored pixels where, in an analog celluloid film, hairs, scratches, and other more organic impurities would appear. The colored pixels reveal right away that we are seeing a digital projection of some sort. But the digital image is a film projection. The screen upon which the film is projected shows soft curls in its fabric, adding to the materiality of the medium. Vertical lines rain on the screen, underlying the horizontal trajectory of the film that twirls in the reels, as the projection machine is audible in the room. The pixels, colorfully peppering the white surface of the screen, seem to be part of a fragmented grid, much of which is missing.

Crossing the line between media has to be considered with reference to line-crossing between translations, but also to the lines of colors. What determines the border between media is the "idea" we have of each medium. For example, film or painting each has a particular "medial configuration."[108] Of course, the configuration *between* different media presupposes actual borders opening a space of oscillation.[109] The borders between media are conventionally drawn—in the same way that *Untitled Translation Exercise* explores the conventionality of language and *Colors* explores the conventionality of color codes. Here, Rajewski is suggesting that the very separation between different media is itself a conventionally defined language or system of signification. The system of references to one medium in another can only approximate an illusion of the original media. Of course, "an overall actualization or realization of the other medial system is impossible."[110] Translation of the original is impossible. A rift opens.

Structural Film is not fully film nor digital projection. It is a film about the degree zero of digital film-making since we already have a content, albeit more or less invisible: iMovie is the content or subject; the narrative story is that something happened to the iMovie filters. Intermediation allows for a reflection about the nature of each discreet media and the conventions determining limits. Intermedia exploration functions therefore like de Man's notion of critical translation: not only does it self-reflexively question the idea of media limits, but it also questions the very nature of the media itself.[111] If the translation was once seen as pointing toward the impossibility of a translation, then intermediality points toward a conventionally constructed nature of each medium: "In other words, they [media practices] necessarily constitute themselves in relation to, and within the scope of, the overall medial and discursive landscape at a given point in time, including the respective delimitations of conventionally distinct art forms and media."[112] The border is a way of looking at the limits within any given medium and the role it plays within an intermedial work or event. But the medium itself, as Rajewsky has demonstrated, is a fluid, abstract, and indeterminate object that is made manifest through a series of conventions: "The concept of a border is the precondition for techniques of crossing or challenging, dissolving or emphasizing medial boundaries, which can consequently be experienced and reflected on *as* constructs and conventions."[113]

What Arcangel's work actually questions is this border—against the conventional. Adding to *Untitled Translation Exercise*'s examination of the conventions of language and to *Colors*' extreme treatment of the conventions of visual semiotics at play in society, *Structural Film* is about the conventional borders between media. But what Arcangel opens is a space between media borders that are not simple lines drawn in the sand between one medium and another that can be breached for cross-pollination. Rather, he uncovers a dimension allowing critical works to independently and fully exist: "The borders or—perhaps better—'border zones' between media can thus be understood as enabling structures, as spaces in which we can test and experiment with a plethora of different strategies."[114] For Rajewsky, the border should be treated as border zones—a space of experimentation, an "enabling structure," just like Deleuze's zone of indiscernibility. Deleuze, after all, already located the "zone of indiscernibility in the line."[115] And here, this Deleuzean concept is what stands between media, but is also what defines the digital and analog through each other's parameters.

Conclusion

The zone of indiscernibility is the rift that opens up between the analog medium of film and Arcangel's digital art. The zone of indiscernibility is a term that Deleuze gave his diagram: "Thus the diagram acted by imposing a zone of objective indiscernibility or indeterminability between two forms, one of which was no longer, and the other, not yet: it destroys the figuration of the first and neutralizes that of the second."[116] From an analog medium like celluloid film (which is no longer in Arcangel's works since it has been appropriated but modified) to a digital one (which is not yet free of analogy since it is heavily dependent on the source material), the zone of passage between the two media opens up and offers a place where each medium can mask itself as the other. This masking of a medium within the traits of another medium brings to mind Rajewsky's photorealist painting and, in turn, the map that is created between two heterogeneous milieus in Deleuze and Guattari's orchid/wasp duality.

In effect, the zone of indiscernibility between the analog and the digital functions (metaphorically) as does the addition of information in modulation. They are placed in contrast with the codification-bound digital. The digital functions through "conversion–translation."[117] The modulation here has to do with the degree of information that separates the two media and the blurring that occurs in *Structural Film*. *Untitled Film Project* started with translation, literal if idiosyncratically redundant, and opened a rift under the pressure of Benjamin, de Man, and Miller to a stratification of sorts. The stratification of interpretations of the theorists themselves created a methodological map of a theory translation. *Colors* nullified the very idea of meaning and introduced an aesthetics of code—artistically, socially, and philosophically deepening the rift between the source object and its modified result.

Notes

1 Dara Birnbaum and Cory Arcangel, "Do It 2," *Artforum International* 7 (2009): 198.
2 Ibid., 197.
3 Rosalind Krauss, "Stieglitz/'Equivalents,'" *October* 11 (1979): 129–40.
4 Birnbaum and Arcangel, "Do It 2," 193.

5 See Robert Burgoyne, "Customising Pleasure: 'Super Mario Clouds' and the John Ford Sky," in *Cinephilia in the Age of Digital Reproduction*, ed. Scott Balcerzak and Jason Sperb (London: Wallflower Press, 2009).

6 David S. Ferris, *The Cambridge Introduction to Walter Benjamin* (Cambridge: Cambridge University Press, 2008), 62.

7 Cory Arcangel, "Untitled Translation Exercise," available online: http://www.coryarcangel.com/things-i-made/2006-002-untitled-translation-exercise (accessed April 12, 2016).

8 Cory Arcangel, "Recent Experiments in Modern Composition, Software, and Stand-Up Comedy," The Art, Technology and Culture Colloquium (UC Berkeley's Centre for New Media), audio recording, available online: http://atc.berkeley.edu/lectures/mp3s/Cory%20Arcangel.mp3.

9 Harmut Winkler, "Computer and the Audiovisual: Ruptures and Continuities in Media History," in *ImageScapes: Studies in Intermediality*, ed. Christian J. Emden et al. (Oxford: Peter Lang, 2010), 231.

10 One would be tempted here to compare art ordered remotely and produced by a call center to László Moholy-Nagy's *EM 2 (Telephone Picture)* (1923). Moholy-Nagy placed a call to a factory, provided the elements of a geometric abstraction using graph paper, and ordered an enamel painting produced to his specifications. Whether this took place (his widow contests there was an actual telephone involved), the remote, anonymous, industrial, mechanical, technological elements of the work are nevertheless resonant. This certainly would play into questions of the materiality of a work of art that starts with the idea, the process, and ends in the hands of a worker on a workshop floor all in the scope of the logic of commodity production and the consequent alienation. But the telecommunication aspect of Moholy-Nagy's work also introduces another type of information age materiality. The artist becomes an "operator of feeds and feedback." See Louis Kaplan, *László Moholy-Nagy: Biographical Writings* (Amherst: Duke University Press, 1995), 125. We are slipping here closer to an information age materiality, negotiating in twenty-first-century call centers between data and non-data selves at the level of information theory, where the boundary between matter and information is "permeable." See Timothy Morton, "Ecology as Text, Text as Ecology," *The Oxford Literary Review* 32, no. 1 (2010): 4.

11 Kalindi Vora, *Life Support: Biocapital and the New History of Outsourced Labor* (Minneapolis: University of Minnesota Press, 2015), 43.

12 Ibid., 44.

13 Ibid., 45.

14 Ibid.

15 Ibid., 46.

16 Ibid.

17 Ibid., 49.

18 Lisson Gallery, *Lisson Presents 7*, available online: http://www.lissongallery.com/exhibitions/lisson-presents-7 (accessed April 13, 2016).

19 Vora, *Life Support*, 52–53.

20 Ibid., 53.

21 Ibid.

22 Ibid., 54–55.

23 In the shadow of materialist exchange between capital and labor is the exchange of a data self and a non-data self. This type of materiality as exchange between biology and information is key to understanding the intermediality of translation between content and form. The notion of individual is problematized through the disembodiment of the non-data self and its embodiment into a data self, and the focus shifts from materialism as exchange to a feed-feedback operating exchange ushered into the algorithmic age by the idea of telecommunication, the seed of which was already sprouting from Moholy-Nagy's telephone picture: "An algorithm is a script—a text—that automates a function, or functions, and in this case the script is encoded directly into matter. The matter-information boundary is permeable" (Morton, "Ecology as Text," 4–5, quoted in Tom Cohen, "Toxic Assets: de Man's Remains and the Ecocatastrophic Imaginary (an American Fable)," in *Theory and the Disappearing Future*, ed. Tom Cohen, Claire Colebrook, and J. Hillis Miller (London: Routledge, 2012), 116). Matter, in this case, takes on the abstract character articulated by Deleuze and Guattari.

24 Brian Massumi, "The Autonomy of Affect," in *Parables for the Virtual: Movement, Affect, Sensation* (Durham: Duke University Press, 2002), 24.

25 Ibid., 25.

26 Ibid., 26.

27 Walter Benjamin, "The Task of the Translator: An Introduction to the Translation of Baudelaire's *Tableaux parisiens*," in *Illuminations: Essays and Reflections*, ed. Hannah Arendt, trans. Harry Zohn (New York: Schocken Books, 2007), 69.

28 Ibid.

29 Ibid.

30 Birnbaum and Arcangel, "Do It 2," 198.

31 J. Hillis Miller, "Paul de Man at Work: In These Bad Days, What Good is an Archive?" in *Theory and the Disappearing Future: On de Man, On Benjamin*, ed. Tom Cohen et al. (London: Routledge, 2012), 79.

32 Ibid., 87.

33 Paul de Man, "Conclusions: Walter Benjamin's 'The Task of the Translator,'" in *The Resistance to Theory* (Minneapolis: University of Minnesota Press, 1986), 81–82.

34 Ibid., 82.

35 Ibid.

36 Ibid., 84.

37 Ibid.

38 Arthur Danto, "The Artworld," *The Journal of Philosophy* 19 (1964): 572.

39 De Man, "Conclusions," 84.

40 Ibid.

41 Ibid.

42 Benjamin, "Task of the Translator," 78.

43 Martin Heidegger, "The Origin of the Work of Art," in *Poetry, Language, Thought*, trans. Albert Hofstadter (New York: Perennial Classic, 2001), 23.

44 Ibid., 23.

45 Benjamin, "Task of the Translator," 82.

46 Deleuze and Guattari, *Thousand Plateaus*, 62.

47 Ibid. Eugene B. Young explains what is at stake in Deleuze and Guattari's notion of matter. First, the link between a language-based poststructuralist materiality as purported by Morton is introduced through the notion of "amorphous mass" explained by Hjelmslev. Matter is linked to the idea of stratification, or rather the non-stratified—matter is "unformed, amorphous, or formless" with its reconfiguration of a disembodied hylemorphic model seen in terms of the abstract machine (another way to say "art" by Deleuze and Guattari) as pure "matter-function." This relates to the cartographic or diagrammatic spatialization of materiality insofar as pure matter lies "outside" of stratification. See Eugene B. Young, "Matter," in *The Deleuze and Guattari Dictionary*, ed. Eugene B. Young, Gary Genosko, and Janell Watson (London: Bloomsbury, 2013), 190.

48 Maria-Christina Villaseñor, "Spectral Projections: Color, Race, and Abstraction in the Moving Image," in *Abstract Video: The Moving Image in Contemporary Art*, ed. Kate Mondloch and Gabrielle Jennings (Oakland: University of California Press, 2015), 198.

49 Camille de Singly, *Guido Molinari: peintre moderniste canadien: Les espaces de la carrière* (Paris: L'Harmattan, 2004), 150.

50 Massumi, "Autonomy of Affect," 185.

51 de Singly, *Guido Molinari*, 150.

52 Robert Welsh, "Molinari and the Science of Color and Line," *RACAR* 1 (1978): 17.

53 Deleuze, *Francis Bacon*, 117.

54 Ibid.

55 Charles Biederman, *Art as the Evolution of Visual Knowledge* (Red Wing: Charles Biederman, 1948), 65–66.

56 Ibid., 65.

57 Charles Biederman, "The Real and the Mystic in Art and Science," in *Theories and Documents of Contemporary Art: A Sourcebook of Artists' Writings*, ed. Kristine Stiles and Peter Selz (Berkeley: University of California Press, 1996), 82.
58 Ibid., 84.
59 Welsh, "Molinari and Science," 16.
60 Linda G. Elson, *Paradox Lost: A Cross-Contextual Definition of Levels of Abstraction* (Cresskill: Hampton Press, 2010), 4.
61 Ibid.
62 Ibid., 5.
63 Ibid., 6.
64 Ibid.
65 Welsh, "Molinari and Science," 18.
66 Ibid. Quoting Jean Piaget and Bärbel Inhelder, *La représentation de l'espace chez l'enfant* (Paris: PUF, 1947), 533.
67 Ibid.
68 Massumi, "Autonomy of Affect," 27.
69 Welsh, "Molinari and Science," 18.
70 Eve Meltzer, *Systems We Have Loved: Conceptual Art, Affect, and the Antihumanist Turn* (Chicago: University of Chicago Press, 2013), 60.
71 Ibid.
72 Fredric Jameson, *The Prison-House of Language: A Critical Account of Structuralism and Russian Formalism* (Princeton: Princeton University Press, 1974), 113.
73 Meltzer, *Systems We Have Loved*, 59.
74 Ibid., 60.
75 Carolyn L. Kane, *Chromatic Algorithms: Synthetic Color, Computer Art, and Aesthetics after Code* (Chicago: University of Chicago Press, 2014), 6.
76 Carolyn L. Kane, "Error as Critical Praxis," *Millennium Film Journal* (Fall 2012): 7.
77 Kane, *Chromatic Algorithms*, 7.
78 Ibid.
79 Ibid.
80 Väliaho, *Biopolitical Screens*, 45.
81 Ibid.
82 Massumi, "Autonomy of Affect," 26.
83 Ibid.
84 Welsh, "Molinari and Science," 10.
85 Stephen Zepke, *Art as Abstract Machine: Ontology and Aesthetics in Deleuze and Guattari* (New York: Routledge, 2005), 207.
86 Young, Genosko, and Watson, *Deleuze and Guattari Dictionary*, 153.
87 Villaseñor, "Spectral Projections," 196.

88 There is a shift between the materiality of the first section (*Untitled Translation Exercise*) and this one (*Structural Film*). Whereas the initial materiality was ideological, a shift began toward a Deleuzean amorphous materiality suited to information. This essential shift is mediated by information theory, which brings everything to the same level (as seen the second section *Colors*). In the present section, we focus on the material, concrete aspect of film stock, but as we will soon see, analogical conditions are enacted by a digital form (and translated again into analog), and a zone of indiscernibility opens up, bringing us back to Deleuze's aesthetic sense of poststructuralist materiality and the cartographic, diagrammatic layout of matter.

89 Cary Archangel's website: http://www.coryarcangel.com/things-i-made/2007-002-structural-film (accessed June 26, 2017).

90 Hanna B. Hölling, *Revisions—Zen for Film* (New York: Bard Graduate Center, 2015), 7.

91 Ibid., 10.

92 Another work by Arcangel, *HITACHI P42H01U Plasma Burn* (2007), deals with the notion of digital decay, as the self-reflexive label information of the work that it is meant to describe is being slowly burned into the plasma screen of a big screen television.

93 Hölling, *Revisions*, 30.

94 Ibid., 81.

95 This brings us back to the idea of the amorphous matter that Deleuze and Guattari have located as the in-between strata. The matter-function of the diagram manifests itself, as we will see shortly, in the zone of indiscernibility.

96 De Man, "Conclusions," 80.

97 Ibid., 83.

98 Ibid.

99 Peter Krapp, *Noise Channels: Glitch and Error in Digital Culture* (Minneapolis: University of Minnesota Press, 2011), 76.

100 Ibid., 91.

101 Ibid.

102 Hölling, *Revisions*, 6–7.

103 Ágnes Pethö, *Cinema and Intermediality: The Passion for the In-Between* (Newcastle upon Tyne: Cambridge Scholars Publishing, 2011), 48.

104 Irina P. Rajewsky, "Border Talks: The Problematic Status of Media Borders in the Current Debate about Intermediality," in *Media Borders, Multimodality and Intermediality*, ed. Lars Elleström (New York: Palgrave MacMillan, 2010), 51.

105 Ibid., 54.

106 Ibid., 58.

107 Ibid.

108 Ibid., 60–61.

109 Ibid., 61.

110 Ibid., 62.

111 Ibid., 63.

112 Ibid., 64.

113 Ibid.

114 Ibid., 65.

115 Deleuze, *Francis Bacon*, 130.

116 Ibid., 157.

117 Ibid., 116.

4

Virtual Images of Swarms and Grids: John F. Simon Jr.'s Posthuman Aesthetics

John F. Simon Jr. animates abstract environments through his algorithm-based art. Algorithmic formulas dictate the aesthetic look of the work: the code and the image constitute a duality that can be described as virtual and actual. Simon's environments serve as a test site for the philosophical notion of virtuality and whether it is suited for the description of digital images. Whereas Simon's digital images depicting abstract spaces seem to fit into an aesthetic continuum of modernist abstraction, they are nevertheless essentially different from images from the past that depend on some form of traditional physicality. To begin, I will establish a definition of information-dependent digital art, a definition revolving around the duality of the virtual and the actual. This definition is constructed through the theories of Laura U. Marks's algorithmic aesthetics; Deleuze's own description of the unrepresentable concept of virtuality; and Christine Buci-Glucksmann's notion of virtuality as part of a posthuman screenic interface. Then, I will look at three different aesthetic spaces: animal, environmental, and architectural. First, *Swarms* (2002)—a metaphorical visualization of the movement of thought modeled on herding, flocking, and swarming patterns—will serve as a model for the animal-image. Then, I will examine *aLife* (2003), consisting of diagrammatic models of ever-changing virtual environments, as an environmental-image. Finally, illustrating the architecture-image, *ComplexCity* (2000) is a representation of Manhattan's gridiron digitally merged with a Mondrian-like map, articulating an impossible space: flat, illusionistic, schematic. These environments are not meant to be experienced by the upright, gravity-bound body that has dictated the orientation of representational art since the perfection of perspective in the Renaissance. Rather, they are made of patterns, schemas, and stylized data that require Deleuze's concept of the fold, Simondon's biologically determined nonvisual notions of the image, and Krauss's art historical artifact of the grid to fully inhabit. These three thinkers, each in their own way, advance an aesthetic theory of space as an image of thought. Deleuze sees space as folds within folds; Simondon postulates nonvisual images independent of representation that are based on pre-perceptive patterns and biological mechanisms; and Krauss sees modernist grids as an actualization of latent scientific theory. The thread will be the actual/virtual duality in Simon's work and how it can be expanded to the nature of digital art.

Digital art and the virtual

But first, to start this chapter, I will need to find a definition of digital art and the virtual. Particularly, I want to look at the one provided by Laura U. Marks and the resulting semantic field, before turning to explore some interpretations of the virtual/actual duality advanced by Deleuze and his critics. The philosophical reading of this duality provides ways of looking at art. More specifically, when it comes to the artistic realm, I will focus on Christine Buci-Glucksmann's interpretation of Deleuze's virtual, which she adapts for the digital age and its relationship to the screen. Furthermore, the notion of swarms will be enmeshed with the philosophical virtual and actual. Finally, I will relate the metaphorical notion of swarms to Deleuze's baroque folds—the baroque, after all, is where the virtual and the actual lie dormant—to find an apt image for the virtual/actual duality fit for digital art.

Marks defines contemporary aniconic art—art that is non-illusionistic, that does not feature recognizable icons, but patterns rather than figures—as information-dependent art. This is a category of art to which Simon belongs, as the work I am discussing here does not represent figures or conventionally recognizable landscapes but deals with moving patterns in an abstract space. Since digital art is, according to Marks, aniconic and therefore does not provide an image of a real object, one might be tempted to define it as abstraction that negates illusion. However, Marks's definition of aniconic art is not constructed in relation to an image or the rejection thereof, but rather around an implicit set of information. This information presents itself in the form of an algorithm or data. She writes: "The image is a selective unfolding of implicit information, and information is in turn a selective unfolding of implicit experience."[1] It is the way this information is made manifest that is intriguing here. Information is folded and unfolded. The fold is, of course, a concept that Deleuze in his book on Leibniz and the baroque explained as an all-permeating mechanism. Marks reminds us that folds carpet the plane of immanence—another fundamental concept for Deleuze. The plane of immanence is also significant in its spatial quality because it is described as "a vast surface composed of an infinite number of folds."[2] Marks explains it as the surface that negotiates between the virtual and the actual: "The plane of immanence is the infinite: it contains all that has existed, will exist, has never existed, and will never exist, in a virtual state. Sometimes one of these enfolded units unfolds and becomes actual."[3] The fold is at the cusp of the virtual and the actual.[4] Marks's own definition of folding of information in digital art depends a great deal on the virtual/actual divide. She writes: "It may be added that what is unfolded into information or image can be considered actual, while what remains enfolded remains virtual."[5] We could repeat and summarize at this stage by saying that digital art takes up the categories of the virtual and the actual and intertwines them with the notions related to folding, and that, finally, these different terms explain the relationship between information—be it data, algorithms, or any type of coding—and a resulting image. This image is, of course, aniconic so as to express the abstract nature of information. But it must be said that the use of virtual and actual here will remain at the level of metaphor. After all, the virtual is not accessible from the actual, and whatever the level of abstraction of a non-illusory image, we are still within the realm of actual representation.

On the one hand the virtual/actual can be seen as a duality: like two different sides of a coin, something that can be flipped from one side to another. Or rather, it can be perceived as something that is unseen, something in the dark and about to be illuminated. According to James Williams, the structure of the virtual/actual has two sides, which is probably why it appeals to Marks as a way of looking at the seen and unseen parts of the digital image. This structure, writes Williams, "is a two-sided one,"[6] where on the one hand we find virtual ideas, and on the other, actual things. But he adds that what makes this two-sided structure dynamic is the involvement of intensities bringing "virtual ideas into greater clarity and obscurity."[7] Even though two-sided, a complication is involved in the structure, which is that of intensities, modulating it so that the boundary line separating the two elements is fuzzier than expected, dimming or brightening reality.

It is necessary to address the issue that the virtual/actual might simply be used as a wedge to split reality into two and in this way rubber-stamp old philosophical models.[8] Aden L. Evens addresses these concerns by describing the stereotypical image of the virtual/actual as primordial goo out of which formed objects emerge. One can imagine here Hieronymus Bosch's closed *Garden of Earthly Delights* (1505–15), with shapes and parts of landscapes emerging from a unidentified liquid. Following those lines, it is typically believed that "the virtual is a churning chaos of molecules in miscellaneous lines of flight which, through accidental interactions among molecules, forms temporary stabilities whose impressions on our senses constitute the actual enduring objects and relations of experience."[9] This type of interpretation spatializes and temporalizes the virtual/actual erroneously: "The virtual is thus a kind of original cause of the actual, and the actual an effect of the *true* events percolating underneath it."[10] The virtual comes before the actual; the virtual is below the actual.

Of course, the relationship between these two categories is not that simple. As Evens explains, organization (organs, organisms, or organization—that is, the whole semantic constellation) does not emerge out of the virtual. Rather, it "tends to cluster and distribute, compress and rarefy, contort and erode, approximate and probabilize. The promise of the virtual is to avoid hard and fast promises, which does not preclude sudden events and radical breaks."[11] This productive imprecision, this malleable spontaneity, is possible because virtual ideas can be considered as traits that are not completely formed, awaiting to be shaped: "Ideas as virtual are fully real though not actual; they exist as structure, as impetus, as incipience, as variation, prior to the individuated world of specific qualities and extended parts."[12] And then the relation between these ideas is what is incarnated in the actual: not the ideas themselves, but inchoate structures, sketches of potential. This transformative essence behind the virtual/actual model is what makes it so appealing to aesthetics.

Following the obscurity/clarity and below/above fallacies of the virtual/actual duality comes the problematization of the boundary. This concept of the boundary is inspired by Kant and explained by Edward Willatt as a space or realm introduced between the virtual and the actual. "This realm secures the conditions of actual situations, it makes the virtual relevant to the actual while preserving and realizing their difference."[13] What Willatt advances is this realm of idea-structures described by Evens, which resembles the schema in Kantian thought. Perhaps with this model we are getting closer at the image of the plane of immanence described by Deleuze as including "both the virtual and its actualization simultaneously, without there being any assignable limit between the two."[14] Far from being a duality split by a boundary, the boundary itself functions as a zone allowing for an interaction between the two. But we are far from the subject of art. How do we bring these notions closer to artistic objects?

One issue that has to be made clear at the level of representation is the virtual/actual spatial organization. This is not a grouping of delimited categories, clearly defined by a surface line below which lies the dark tumultuous water and above which is the daylight of the real that makes things recognizable. Robert Motherwell's description of Jackson Pollock's take on Pablo Picasso is clarifying in this respect, in its complication of the duality's organization.[15] Imagining Pollock's abstractions as the surface of a water teeming with life, Motherwell described how the heavy-drinking artist, feeling that the figures he painted in the style of Picasso were inadequate, would violently cover them with slashes of his paint brush. This image stayed with Motherwell, who would see Pollock's paintings as the surface of a less abstract but murky depth, a marshland of sorts. But let us shift this immersion surface/depth metaphor to a discussion about the philosophical virtual and actual. And in order to once again complicate the images of clarity/obscurity, below/above, and the divisive boundary, Evens adds a little bit of liquid to this summarizing description: "Ideas flicker and swarm at the edges of the subject, ensuring and problematizing its engagement with the world by immersing its borders in the virtual."[16] Instead of a division in two separate realms—one below or beyond, the other above or within reach—we have a dynamic complication of these dimensions through an immersion, as if dipping but not wholly swimming in the primary goo mentioned previously.

If modernist painting can offer a model for thinking about the division of the virtual/actual's false duality, can the virtual/actual in fact be used positively to illustrate the parameters of the digital image? First, I will examine how Deleuze describes the virtual/actual; then I will consider how it can be used in aesthetics; and finally, how it can be harnessed, via Leibniz, as a metaphorical way of thinking about the digital image.

Deleuze's image for the virtual/actual duality is spatial in a deep-space manner of speaking—a core body surrounded by orbits that are nevertheless in constant motion:

> The virtuals, encircling the actual, perpetually renew themselves by emitting yet others, with which they are in turn surrounded and which go on in turn to react upon the actual: "in the heart of the cloud of the virtual there is a virtual of yet higher order … every virtual particle surrounds itself with a virtual cosmos and each in its turn does likewise indefinitely."[17]

Deleuze starts with a singular perception, a perception-particle: one single unit of sensory-motor awareness, clouded with a multiplicity mist of inchoate potential images.

In the same essay, Deleuze makes the relation between the virtual/actual a little less abstract by turning to an Orson Welles film to illustrate the function of the terms. The example he chooses is a film sequence from *The Lady from Shanghai* (1947) that puts into motion a series of mirror images. If Deleuze has elsewhere explained that the mirror image in film is the virtual of an actual character it captures, this becomes more complicated in a scene in which three characters enter a hall of mirrors, and two of them, only barely, come out:[18]

> In *The Lady from Shanghai* … the mirror takes control of a character, engulfs him and leaves him as just a virtuality; hence, there is coalescence and division, or rather oscillation, a perpetual exchange between the actual object and its virtual image: the virtual image never stops becoming actual. The virtual image absorbs all of a character's actuality at the same time as the actual character is no more than a virtuality. This perpetual exchange between the virtual and the actual is what defines a crystal.[19]

A veritable cinematic machine, Welles's famous hall of mirrors scene plays out a series of accusatory gazes, the underlying film noir assumption that things are not as they seem, with a series of mirrors that look very much like an inhabitable crystal that is in constant motion. The mirror takes control of the character because it becomes an agent in the action, determining the character's fate.[20] And time—albeit a time that is beyond the human experience of it—is reflected in the virtual image of the mirror. Perhaps a succinct analog of this mirror scene is Rodney Graham's *Torqued Chandelier* (2005), a spinning core deterritorializing the decorative crystals into a dazzling optical machine. "Inspired by Sir Isaac Newton's famous water-bucket experiment, which explored the nature of rotational motion, *Torqued Chandelier Release* documents a crystal chandelier—wound up on a rope off-camera and then released—spinning in one direction until the rope unwinds, slowing, then spinning in the reverse direction, and so on, until finally coming to rest."[21] A relationship can be drawn (and has been) between the luminosity of the chandelier as a metaphor for the enlightening experiment perpetrated in the baroque era—and, of course, by Newton, Leibniz's contemporary—and as a visual metaphor of illumination of the mind.[22] It is a pure crystal, mirroring the light that comes straight at it. And it comes out again, illuminating the dark room housing the projector and the screen: a tautology of the crystal concept: the black background, the white crystal, like a jewel in a black velvet box.

Resorting to yet another image would perhaps clarify the matter somewhat—or perhaps make it more opaque, as we are moving from crystal to marble. But just like the crystal and its virtual dynamism, the marble, as inorganic matter, is teeming with motion. Deleuze sees the inorganic matter of a block of marble as a fish pond (reminding us of a Pollock painting):

> It is a lake, a pond, or a fish hatchery. Here the figure of the lake or pond acquires a new meaning, since the pond—and the marble tile—no longer refer to elastic waves that swim through them like inorganic folds, but to fish that inhabit them like organic folds. And in life itself the inner sites contained are even more hatcheries full of other fish: a "swarm."[23]

How does this "swarm" give rise to ideas and make inorganic matter living? Christian Kerslake, who studied the figure of the marble block in light of the actual and virtual division, can clarify the matter. He explains that marble veins stand in for the formation of ideas in Leibniz and by extension Deleuze. He writes, "Veins and shapes are present on the surface, as well as within the depths of the marble, but these can only be discovered by being chiseled at and exposed to the light."[24] In this process, some come to the surface while others remain latent. So the shape of Hercules, to follow Leibniz's example, is virtually outlined in the block and will come out, not by its own volition but because it will be nudged out. The posthuman lack of agency is there, firmly embedded in the concept of the virtual: the swarms will continue to form, the flux-image on the surface of the screen will perpetually continue to animate the algorithm that determined it, whether we are watching or not.

Some of these elements are present in Buci-Glucksmann's theories of the virtual. In *Modèles du virtuel*, she sees virtual technologies as an archaeological ground that needs to be dug up so as to excavate traces of the posthuman. Why call it the posthuman? Because the time in which humans usually orient themselves is no longer straightforward. Rather, time is machinic and mechanized, ephemeral and yet seeming eternally present. Buci-Glucksmann deals with an art of the virtual that no longer appeals to a subjective, essential experience. Rather, it seems to disrupt our expectation vis-à-vis the biological, the ethical, and political, as she claims. The time of new technologies offers something nonhuman: it does not operate according to the biological rhythms ingrained in the human body. The human body is itself split between the real and the screenic virtual (*virtuel écranique*).[25] A good example of this is Walter Benjamin's notion of the actor's aura: actors are losing their aura because they no longer perform plays in front of an audience where they can gauge reactions and respond organically; instead, their biological rhythms are chopped up and reshuffled in front of the camera and the sequence is only put together in the editing room.[26] Orson Welles and Rita Hayworth, the two stars of *The Lady from Shanghai*, are chopped up and multiplied through the hall of mirrors like an illustration of the modernist loss of aura in Benjamin. Whether behind the camera or in front of a mirror, the modernist image's relation to time is at a standstill compared to what is to come with the digital manipulation of time. Buci-Glucksmann summons the actual and virtual to energize Deleuze's treatment of static modernist images (which Deleuze sees as a crystal with hard edges, no less) into flux–images of the digital, posthuman world. Maybe Benjamin, Hayworth, and Welles miss the mark—they are too far back. Maybe Buci-Glucksmann is thinking of someone hunched over a softly glowing computer screen.

But Marks had already located this shift from sensuous experience to informational process in Deleuze's own prediction about the changing nature of the screen, which, "even if it keeps a vertical position by convention, no longer seems to refer to the human posture, like a window or a painting, but rather constitutes a table of information, an opaque surface on which are inscribed 'data,' information replacing nature, and the brain-city … replacing the eyes of nature."[27] Marks reads this as a shift from visual to information culture, from electronic to digital images, from a "window out to the perceptible" to one that is "in to the legible."[28] The screen is not the silver screen but rather the opaque computer screen, and not the retinal computer screen through which we experience contemporary media, which can be as illusionistic as cinema of the past, but the green type flickering on a dark background in *Alien* (1979) (MU-TH-UR), or *War Games* (1983) (WOPR), or *Space Odyssey* (1968) (HAL)—bearing witness to a virtual out of reach that was merely informing us about the environment over which we have no control.

This screen has been presented to us by Leo Steinberg, of course, and his theory of the flatbed picture press as a passage from nature to culture. Deleuze takes up this notion again in *The Fold* to qualify modern baroque art (Simon Hantaï, a French abstract painter who worked with folds and grids; and Rauschenberg, who seemed to create canvasses upon which the ephemeral of the television age is stuck) as an opaque grid of information.[29] Now, suffice it to say, the virtual/actual divide is an apt tool for both Marks and Buci-Glucksmann—and even, indirectly, Deleuze—to look at information-based art.

What seems to be emerging in this discussion of the actual/virtual in the domain of art is the notion of the posthuman. This is why we will turn toward the image of the swarm as a fitting one through which to consider it. We are not discussing here an impersonal mass intelligence, in the way, for example, that the Borgs on *Star Trek* function as a hive consciousness, but are rather looking at how animal patterns can be seen as post-individualistic in this information age.[30]

Swarms

"What is the essence of the swarm?," asks Alexander Galloway in his article on the video game StarCraft. His definition is multisensory, both acorporeal and multicorporeal, lacking a head and a humanity:

> The swarm is a resounding-forth. It is a buzzing, an articulate, identifiable murmur that is nevertheless unendowed with an emergent spirit or soul. If a resounding-forth is the essence of the swarm, it is a resounding-forth that cannot be further reduced to a hermetic, singular ego. It is acephalous and unhuman. It is a disavowal of centering, of genetic reduction to any sovereign essence. This is what it means to say that the swarm is unhuman. Certainly the swarm has as its "essence" the creation of presences, as in the resounding-forth. But the swarm also brings about the destruction of that uniquely "human" form of presence.[31]

This is an excellent definition of the type of posthuman swarms that we will be exploring through digital art. Simon's *Swarms* [*Figure 10*] both illustrate this idea of the posthuman aesthetic of the screen. Simon describes his *Swarms* as a "software program running simultaneously … across two 50-inch plasma monitors" that make up a singular window.[32] The unusual arrangement of the monitors brings attention to the screen experience of the viewer. He writes that it is the "software that shows the formation and dissolution of symmetric patterns as a metaphor for the temporality of our thoughts, ideas and actions."[33] Just as predicted by Buci-Glucksmann, the screen and software here are in cahoots with time and its effects on thought—which is part of the posthuman thread being pulled out of the virtual/actual duality. Nine swarms of patterns of various color combinations gather and disperse in formations that will never repeat. But the title, *Swarms*, also plays a role in not only being a cloud of images in Deleuze's text, but the mass movement of beyond the human multitudes.

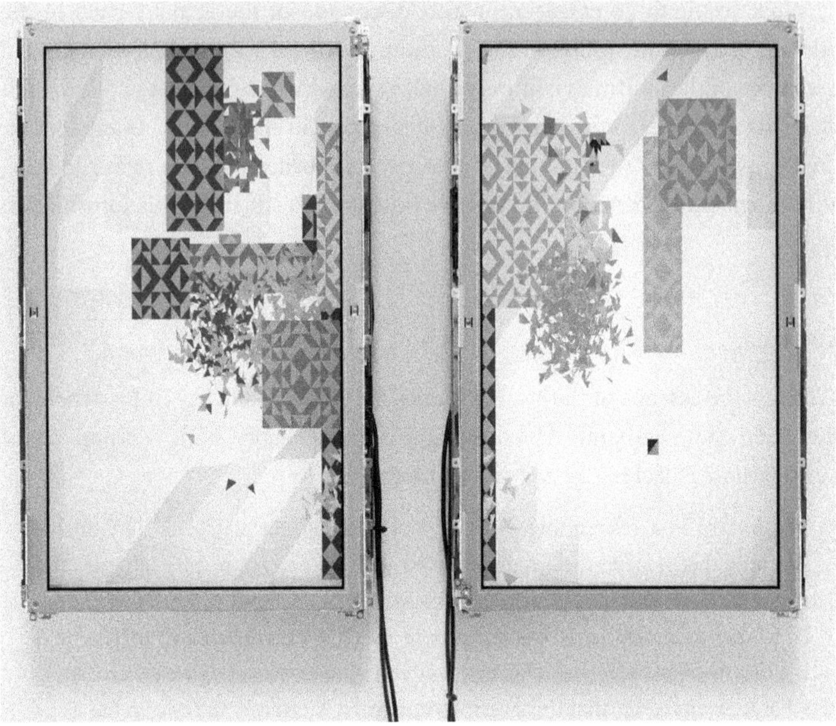

Figure 10 John F. Simon Jr., *Swarms*, 2002, Software, Macintosh G4, two screens, runs continuously, never repeats. Courtesy of the artist.

For this work, Simon created an algorithm to produce these flowing designs. Outside of the domain of art, algorithms based on swarms have been developed in the field of computational intelligence: computational methodologies designed to address real-world problems of avowed complexity that take their inspiration from patterns found in nature.[34] Swarm patterns found in schools of fish or flocks of birds—technically referred to as particle swarm optimization (PSO), where the particle refers to the individual of a flock—are used for their cooperative ability to function as a whole and find food or sanctuary in large spaces previously unknown to the swarm. This function is then translated mathematically to solve engineering problems. But if the swarming movements witnessed in PSO are abstracted from actual models, what Simon is doing is creating an algorithm to give the visual effect of swarming a virtual dimension.[35]

Somewhere between the art made by Simon and the science of PSO is *Swarmart* (2004)—an equally important piece on the subject of animal patterns that I would be remiss not to mention. This collaboration between Gerald Hushlak and members of the computer engineering department of the University of Calgary produces swarms that are activated by spectators moving in front of a screen. The human aspect is vital here, however, and it appeals more to a physicality, a verticality. It is precisely what Simon's swarms do not have that lets us see how integral the conception of a virtual is to digital art and how it eschews the necessity of the human body. *Swarms* is "horizontal": we could be seeing caribou herds moving through the tundra. *Swarmart* is "vertical": the body of the viewer is made to interact with an illusion of space:

> The evolutionary approach proved to be very useful, as the choreographer/designer does not require any programming knowledge. Instead, the designer "breeds" swarm behaviors using a graphical interface through which the dynamic swarm patterns are visualized and evaluated. The GP system takes care of mutations and recombination of simulation parameters, depending on the breeder's selection preferences.[36]

But because of its dependence on gravity, on the upright body and use of direct analogies of movement, this piece is still closer to modernist anthropocentrism.

How are the swarms roaming the screens of Simon's art piece different from Hushlak's *Swarmart*? They are units of geometric patterns made of similar shapes, breaking up and coming together. Metaphorically speaking, the geometric patterns seem to behave like groups of animals. The two screens break up the unity of a single surface resembling an abstract painting. They reinstate through the gap between the screens another space not contingent on the frame—which is a denial and expression of the very modernist tenets advocated by Clement Greenberg. Even though the work is abstract and colorful, the frame is breached, incidental; and yet from a media point of view, the breach reaffirms the frame as screen. While simultaneously breaking away from a space of analogy with the surface of painting, the spatial elements are reinstated through the analogy with animal hoard behavior. The strong spatial element of the piece now can be treated as a map or an abstraction of animal movements. Are we looking at a murmuration of a starling flock in a landscape, a schooling of fish in water, a hoarding of caribou seen from above, or the swarming of bees? Under what circumstances would these images constitute maps of the actual and the virtual that, in turn, would self-reflexively illustrate the defining function of digital art?

aLife

Simon's work *aLife* [*Figure 11*] is a virtual environment putting into motion informational graphics: composed of digital map-images of indeterminate worlds aligned with posthuman aesthetics that seem to ask: Who are images for? What if they are for no one? I will be looking at Gilbert Simondon's concept of

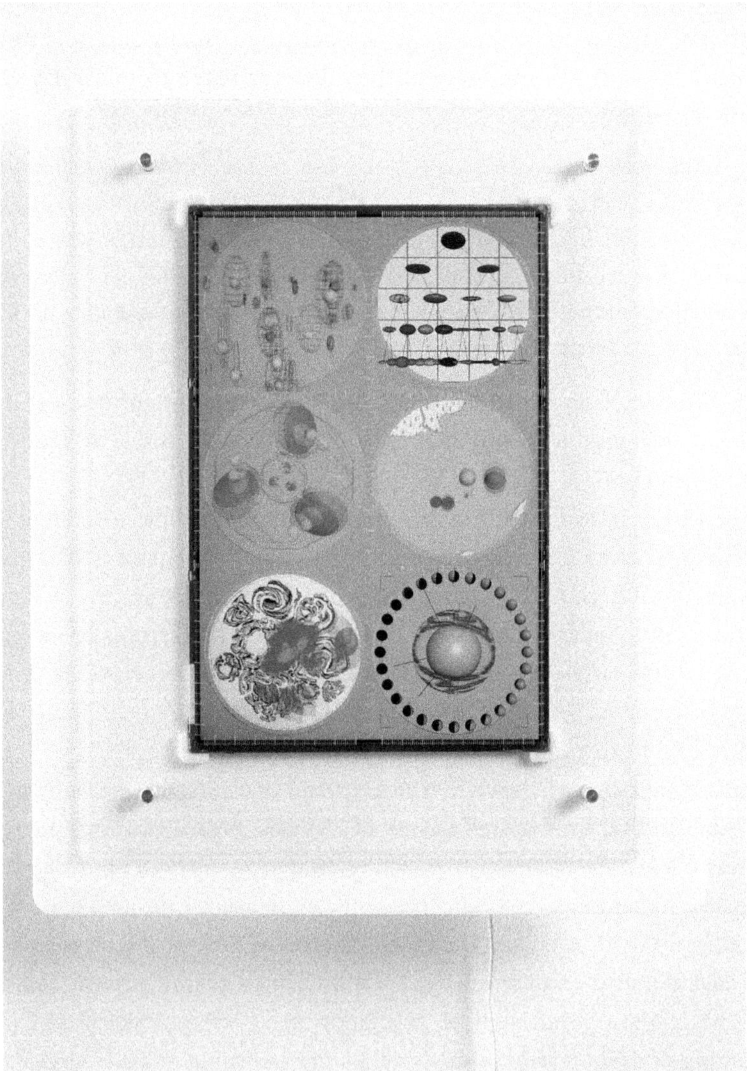

Figure 11 John F. Simon Jr., *aLife*, 2003, Software, Macintosh G4 PowerBook, and plastic acrylic, 21 × 17 × 3 inches, runs continuously, never repeats. Courtesy of the artist.

the intra-perceptive image that scans an environment without actively looking at it. As before, I will use Simon's work to stand as a frontispiece to these notions of a (more or less) posthuman image.

I want to consider Simon's cartographic work because it brings about a posthuman world and illustrates the type of disinterested, or unmotivated, images that must be read according to an actual/virtual duality. Simon's *aLife* is described as offering a "glimpse into the evolution of six tiny worlds."[37] But what is interesting is the way that the worlds have been formed: each world resembles the schematic designs of the scientific discipline dedicated to it. So, the world of weather is made up of images seen in meteorological satellite maps. We know we are looking at an atomic world because we see the patterns usually used to depict molecules. Visual information used to depict worlds that are not visible to the eye is then appropriated as the signage of a particular field to create diagrammatic worlds perpetually evolving and changing. It is as if conventional signs have taken a life of their own, independent of their deciphering eye. What is more, what Simon seems to be staging here are worlds beyond human perception. These are nonhuman-scaled landscapes. In this work, Simon seems to tell us about the limits of perception and the way visuality has to be reconsidered in the digital age. The shift between information about an environment and the skin this information must wear in order to be recognizable are cleverly depicted here.

A close analog to Simon's work would be John Klima's *Ecosystm* (2000). Here, data is a recognizable source at the root of the motion, and the viewer is encouraged to decipher changes animating the world on the screen in the manner of a mariner divining the sea. Again, Klima is a straw man here: this comparison highlights the posthuman aspect of Simon's work.

> *Ecosystm*'s display consists of virtual, bird-like creatures whose movement occurs in relation to the value of different currencies, and the work includes a joystick that allows the viewer to navigate the environment, its terrain dotted with trees representing global stock-market indexes. The real-time currency data that animates the flocks of birds comes from the CNN web site, rendering a continuous visual picture of global capitalism.[38]

What Klima has done is to reshape the abstract financial system as an evolving biological environment. In describing his piece, he mentions a flock occupying a territory that moves according to the volatility of the market: "If, however, the currency is volatile, the flock becomes very 'excited', and their available territory is considerably reduced in size … if the daily volatility exceeds twice the yearly volatility, the flock is 'hungry' and it 'feeds' on its country's leading market index (as represented by the trees)."[39] Klima is staging something beyond human perception: the market. This has been explained by Pierre Lévy in *Qu'est-ce que le virtuel?*, who states, "Contemporary economy is an economy of deterritorialization or of virtualization."[40] Coupling the virtual with another Deleuzoguattarian term, deterritorialization, Lévy describes a world economy where the highest financial activity is concerned with the not-being-there and where everything from currency to GDP is virtual insofar as it is not present and not reachable. This is why Klima's work, which also seems to consider the notion of space and the virtual in relation to the market, creates an environment where virtual values can be apprehended through perception. But what we have here are nonhuman semi-abstract environments. What if the environments just got rid of the human instead?

Jean-Clet Martin, in his essay *Le virtuel incarné*, discusses a type of image that does not fall under the system of human perception. He is writing about the virtual image that is neutral and impersonal, different to some degree from Buci-Glucksmann's posthuman flux-image in the sense that it is less about the biological rhythms of a human squeezed from all sides by technology. It is an image, according to Martin, that crosses several thresholds of perception greater than human perception.[41] Martin's image is posthuman-pastoral—far away from technologies. Martin looks up at the clouds and realizes they are not meant for him; he is not the target audience of the cloud-produced images.

Martin wants to know how we can pick up these images that are beyond human experience: digital images that come to be without a particular point of view. He settles on the example of the cloud. Of course he is discussing the weather, but one's mind rushes to Deleuze's description of the virtual as a cloud surrounding the actual. Here, the cloud is a cloud. Martin says that its form and the organization of its colors do not depend on the logic of the gaze. The cloud, Martin continues, does not need us and does not depend on our point of view to constitute an unmotivated, pure image. If the cloud is to be motivated, it is only by the wind, of which it is the negative, dynamic image.[42] Martin believes that only the virtual image can navigate the storms of heterogeneity in the world because it is based on data and not feeling, providing the point of view of the virtuality of the world since the visual is already part of things.[43] Like Martin's unmotivated image, Simondon's concept of the nonvisual image, described in *Imagination et invention*, provides a reading of Simon's posthuman worlds.

Simon's worlds in *aLife*, following Martin's idea of the infographic images and the morphing that they undergo, are spinning, turning, floating (one resembles spherical bodies floating above a map; another seems to be schematized hemoglobin; still another seems to be variation in atmospheric representation).[44] The six configurations of Simon's *aLife* are organized on a grid.[45] Underneath this grid is an algorithm coded by Simon that gives the impression of living evolutionary systems. The source material involved comes from "scanned maps," "scientific diagrams of subatomic structures," "planetary systems or microscopic organisms," and even a nod to modernist design in the form of a George Nelson Ball Clock.[46] But it is the underlying, invisible code that articulates the environmental structures.[47] Here, I would like to look at Simondon's intra-perceptive image—which, like Martin's virtual image of visuality already being part of things, is not the representational image that we have in mind from an artistic or cultural perspective. Like Martin's cloud, more or less, it is an image that we do not think about; it is before our thought. I would like to suggest that Simon's infographic worlds are an illustration of Simondon's intra-perceptive images. And on some level, Simondon's concept of the image is a part of the environment, but as an imperceptible underlying structure.

Intra-perceptive image

Simondon's philosophical system was first published in 1964 as *L'individuation et sa genèse physico-biologique*. It has recently garnered attention because of Simondon's studies in tandem with Gilles Deleuze. Deleuze, who sat on Simondon's doctoral committee, greatly admired his systematic philosophy and incorporated aspects of it into his own system. What I want to explore here is the role of the image beyond representation and how this can be used for contemporary aesthetics. One of the features of Simondon's philosophy is how the image is part and parcel of the process of thought. Take, for example, a passage from *Imagination et invention*, lectures given by Simondon in 1965/66 at the Sorbonne. "The image," Simondon writes, "is a thought process already partially formalized."[48] So how does the intra-perceptive image help us think about aesthetics? According to Jean-Hugues Barthélémy, "Simondon shows that that which *precedes* perception—that is to say, the motricity of the living—is already the birth of a 'cycle of the image' that extends into perception itself in the form of 'intra-perceptive images.'"[49] It would not be a stretch to read this pre-condition of perception as a virtual image. Simondon explains the primordial importance of intra-perceptive images in progressive behaviors, where perception occurs after perceptivo-motor activity and unfolds in a temporal sequence. Simondon gives the example of the predatory beewolf, a species of wasp that preys on bees. The presence of a bee spurs the chase, but the beewolf will only have a clear perception of its actions once the capture is made. Simondon explains how a beewolf reacts and pursues not as a specific individual, but using traits resembling a number of members of several species of bees. All this occurs before it has even caught up to the individual prey. If the beewolf waited until it precisely determined the nature of the blur flying past before engaging in pursuit, its supper would be long gone. Each stage preceding the capture is a sketch of a final perception. All together, these sketches constitute an image.[50] Potential traits, teeming with possibilities at the stage of the sketch, come to a rest in the object.

Seven features of the intra-perceptive image as defined by Simondon emerge: trait, trigger, pattern, code, territory, grid, and, as mentioned above, virtuality. The intra-perceptive image operates by perceiving only traits of an object prior to a final synthesis. It is triggered, meaning that it is not actively sought. It operates through organizational patterns. It functions according to a code—a set of rules that define its apprehension. It is environmentally dependent and is associated with a territory. Its underlying structure is that of the grid—virtually, when, for example, organizing a territory; or actually, when an actual grid pattern is used in an aesthetic object. Finally, it tends to dip into the virtual because it is about potentiality.

Trait. The image is a "configuration, a grouping of traits and not a determined object."[51] We are dealing with traits, and not complete images. Certain traits act as *releasers* of instinctive behavior in all animals. But to illustrate the notion of trait with a familiar example, Simondon notes how the film industry condenses "perceptive configurations" of traits resulting in types—such as the woman-child, like Shirley Temple or Brigitte Bardot. At another time in history, the Venus of Willendorf exemplified a figure made up of exaggerated traits. Traits allow for smaller doses of input and thus greater flexibility during output.

Trigger. The intra-perceptive image also triggers "attitudes," postures, or "information selectors."[52] Simondon clearly underlines the inchoate aspect of these images. It is responsible for the instinctive reaction they elicit. But as a trigger, the image can also respond to a specific configuration—a snake seen by a bird, as Simondon points out. He explains that the triggering mechanism of such an image can be employed in two different instances: the conclusion to an initial series of progressive sketches of an object or the introduction to an activity of consumption. On this trigger seems to hinge a possibility of utilizing the intra-perceptive image aesthetically.

Patterns. Simondon writes: "The intra-perceptive image plays the role of model, the '*pattern*' of larger generalities to which are compared the totality of incidental signals."[53] The image here is a template to be filled out by information. The relationship between information and such a pattern is instrumental since, according to Simondon, this duality allows for a dynamic progression of the image. The pattern complies to a continually inputted set of varying information and thus allows for a calibration on the fly, or as Simondon states, "Received information is compared to this situational '*pattern*' on a continuum."[54] This leads Simondon to describe the image as a "virtual object the apparition of which in a particular place is anticipated from the surrounding conditions."[55] This is crucial—the image is already present in an environment awaiting to be triggered into actuality.

Code. The idea of code depends on the last characteristic of the pattern. Perhaps Simondon's example involving a fighter jet would best serve to explain the coding, or translation, characteristic of the intra-perceptive image. When we hear a fighter jet pass overhead, Simondon explains, we automatically look away from and ahead of the source of the sound: "The intra-perceptive image [is] the lag between sound location and visual location … the image supposes then a code of transformation of the object, a formula of potentiality allowing us to foresee the transformation of received signals in relation to the surroundings and the developing action."[56] The code anchors the intra-perceptive image through habit.

Territory. We have already discussed patterns, space, and animal behavior. Simondon explains next that the image is already in a territory, present in a space, and can concretely, without representation, become manifest. This is interesting since the image can be more than visual, or visual without being seen as such. Some of the examples here are animals fleeing a space they know intimately at the first warning signs of an upcoming cataclysm. This type of perception calls for a singular type of knowledge of the link between all the details of the environment. It is the linking that creates a territory instead of a simple field of activity.[57] Other examples of territory include that of the baseline image shepherds have of their flocks, so that they are able to tell without counting whether animals are missing. Of course, Simondon says, the shepherd does not count all the sheep in his head: A lag appears clearly between the baseline image and the perception data.[58] The lag appearing between the image and data is not perceived as a representation. Another example Simondon uses to describe the notion of territory is *l'esprit de finesse*, one's ease with crowds: "Its operative mode comes out naturally from the perception of the state of things. L'esprit de finesse resides in the individual who already possesses an image of the organism it approaches."[59] This is the individual's aptitude of discerning attitude in a mass of people, here taken to be a territory.

Grid. This notion of territory gives rise to the concept of the grid: the relationship of image to information. Simondon describes an intra-perceptive image that bears witness to the eye's enormous capacity for sensory reception of information as opposed to the mind's attentive apprehension of discrete elements, which is limited to only a couple of binary units per second. Simondon asks why the eye has the capacity to capture millions of distinct points of view in the nick of time. He believes that the potential of this capacity nourishes a differential receptive activity. Take the subject whose eyes perceive the world by processing external data as if it were correcting the normative grid of an exam sheet: it skips the uniform patterns and only registers mistakes. On a serene field of vision, aberrations stick out.

Virtuality. The virtual seems to have already come up in one form or another in the preceding categories: virtual object anticipated by a place rendered actual; the potentiality of each sketch leading up to a representational image; traits as the virtual to an actually formed image. With regard to the grid: geometric forms are a manifestation of the virtual because they provide access to all sorts of possibilities that lie in contrast to an established order of nature. The intra-perceptive image, in fact, offers a portal to the virtual.

But how does the intra-perceptive image affect art? How can we apply the intra-perceptive image to art as a critical device? The different iterations of intra-perceptive images are not symbolic or allegorical; they are not didactic or even beautiful. They are not, as Simondon explains, "*a posteriori* constructions summarizing experience."[60] They are neither contemplated nor reflected upon. They are, on the contrary, "*a priori* facilitating the insertion of a living being into his environment."[61] They are functional and they are spatial. No wonder that the two positive examples of aesthetic manifestations of the intra-perceptive image in the domain of art involve brutalist architecture.

Simondon contrasts two categories of art: analytic and synthetic (neither having anything to do with the early stages of cubism). Synthetic art involves adding—you have a house; you add a painting. You add objects to objects.

Analytic art, clearly the superior category, "does not produce supplementary or secondary objects that mask base, primitive, objects."[62] The material make-up of analytic art is esteemed through its texture and configuration. For example, nothing needs to be added, neither paint nor plaster, to a granite wall: "[The] material already possesses microstructures, an original texture that cutting, polishing and treating it with a sand blaster can help bring it out without hiding or taking anything away."[63]

The first example of a successful intra-perceptive image in the domain of architecture is *La Tour Croulebarbe* (1961), a grid-patterned skyscraper in the thirteenth arrondissement in Paris built in 1961. Simondon's focus is on the metal sheets covering parts of its gridded façade. Noticing the longitudinal folds that come with sheet metal, Simondon explains how the folds should be arranged vertically on a building to give it an effect of height. Of course, the same material applied to a train wagon should be arranged horizontally so as to reinforce the effect of forward movement. His example provides a way of thinking about how the orientation of certain forms is expected because of the rational function they serve. So, a train will have horizontal sheet metal because of aerodynamics, but the same sheet metal will be vertical in a building in order to facilitate rain water drainage. But this aesthetic choice is also functional and works according to the operation of objects. This is what Simondon calls an "intra-perceptive rationality."

The other manifestation of this image is Le Corbusier's Dominican convent at Arbresle. It is a monolithic structure of cement planted in idyllic surroundings. Where one would be fascinated by the austerity and the starkness of the structure, Simondon sees life: "Like a living organism, each line of the total configuration is multifunctional; a corridor is a site of passage; it is a collector and also a distributor not only of human beings into different rooms but also of electricity, air, water and information; what is multifunctional is open and non-saturated."[64] Humans, information, and electricity are on the same level, flushed down the corridors into individual rooms. It is that kind of thinking that led to accusations of coldness, unfeelingness, or even a totalitarian taste to be directed against Le Corbusier by his critics. For Simondon, the structure elicits fuzzy feelings—he believes that this analytical art is the most welcoming of new realities because its image is not material but "bridges texture and configuration."[65] This welcoming functions not with open arms, but rather as a result of its rationality vis-à-vis the material. What impresses Simondon the most is that Arbresle is perhaps the only convent in the world where a metallic propane cistern could abut the chapel and not be perceived as an anomaly. The convent at Arbresle is an example of a coupling—the essence of the intra-perceptive image—between nature and technology.

The other category of art, the synthetic that adds object to another object, is referred to by Simondon as contemporary baroque: it does not contain any traces of coupling or of the intra-perceptive rationality. When one thinks of the baroque, Rembrandt's Chiaroscuro comes to mind, or Caravaggio's high drama, or even Bernini's flowing waves of marbled cloth. What the philosopher refers to with the term "contemporary baroque" is Op art: made popular by Bridget Riley and Victor Vasarely. Op art was in its heyday when Simondon was giving his *Imagination et invention* lectures at the Sorbonne with exhibitions around Paris and a major exhibition at the MOMA in New York in 1965 called *The Responsive Eye*. If the term "baroque" is used in a derogatory way, Op art nevertheless seems to fulfill the condition of the virtual as a portal for potentiality in the relationship between grid and nature. But for Simondon, this type of art separates independent microstructures from configurations: they don't have a functional rationality/relationality, leaving the intra-perceptive image "floating and indeterminate."[66] He writes: "A contemporary form of the baroque appears and develops with the proliferation of automatism of Optical Art, microstructural, geometric and contrasting motifs are developed for themselves and then are used to dress objects that already have their own sense and form."[67] The only redemptive aspect of this contemporary baroque, according to Simondon, is that it is inspired by "*techniques de balisage*"—marking techniques applied to such objects as tracks, roads, targets, and racing flags.[68] However, he does take issue with the fact Op art uses them purely rhetorically.

One element in common to all three works critiqued by Simondon is the grid. This visual trope has been, of course, famously covered by Rosalind Krauss in her essay *Grids*. There she focuses on early twentieth-century modernists such as Malevich and Mondrian. These artists, she states, have closed themselves up in a mute ghetto that pushed away audiences who have an insatiable appetite for narrative. Spectators want to look at a painting and have something to say. Perhaps even something banal, like this is a cup or a bowl of fruit when looking at a Cézanne or Picasso. But the issue with the grid is that it is anti-narrative, anti-diegetic, anti-aesthetic, and against language. Humans, after all, like to tell stories, like to imagine a face in the abstraction, a face in the cloud, even.

Simondon's nonvisual, information-based image does not require interpretation, or an informed gaze, or even agency. It is already present in a disinterested environment. The grid pattern is part of Simondon's explication of unmotivated images: it underscores the idea of perception through abstract patterns. The image is already virtually present in an environment—or a face or a situation—as information, and an event will actualize it.

ComplexCity

The intra-perceptive image as we have just seen it is the combination of the swarm and the grid. But now I will look at how the grid-pattern itself within the context of art—and to a certain extent, art history—is subject to the virtual/actual duality. *ComplexCity* seems to summarize from the point of view of gestalt the virtual/actual duality we have been exploring from the perspective of aesthetics [*Figure 12*]. According to Simon, "*ComplexCity* is concerned with the realism of abstraction."[69] I wish to explore this image for its grid pattern and how it illustrates the notion of actual and virtual. Simon describes his work thusly: "A modern day cityscape, including skylines, skyscrapers, and a traffic flow reminiscent of Mondrian's *Broadway Boogie Woogie*, ComplexCity explores the pure abstraction of Mondrian and the abstract realism of Stuart Davis."[70] Two elements are combined here: gridiron pattern and abstract art. Simon's work focuses on Manhattan—which is based on the gridiron plan—and abstract painting about the city.

While Stuart Davis's cityscapes can be surmised in Simon's piece from a loose perspective, it is Mondrian's encounter with the digital that is really interesting. This is not Mondrian's first encounter with the digital. Mondrian's work was taken up by pioneering computer programmer and artist A. Michael Noll in 1964. The computer program devised by Noll was made to imitate Mondrian's *Composition with Lines* (1917): "Noll varied the degree of randomness, which ranged from a grid-based placement of varying-length lines to a completely random placement."[71] Noll then asked one thousand people to compare Mondrian's work to the computer-made one—the mimicked Mondrian was taken for the real one. Mondrian was at the dawn of computer art: "This and other programs by Noll marked the beginning of computer art as a branch of autonomous arts not a tool for studying existing works."[72] We are applying our knowledge of Simondon's intra-perceptive image in that "every desired property not formalized well is laid bare immediately [in computer art]" and that the computer can then be taken, following this idea of gridding a landscape and waiting for the unexpected object to pop up, "as a tool of careful observation."[73] Of course, this would force us

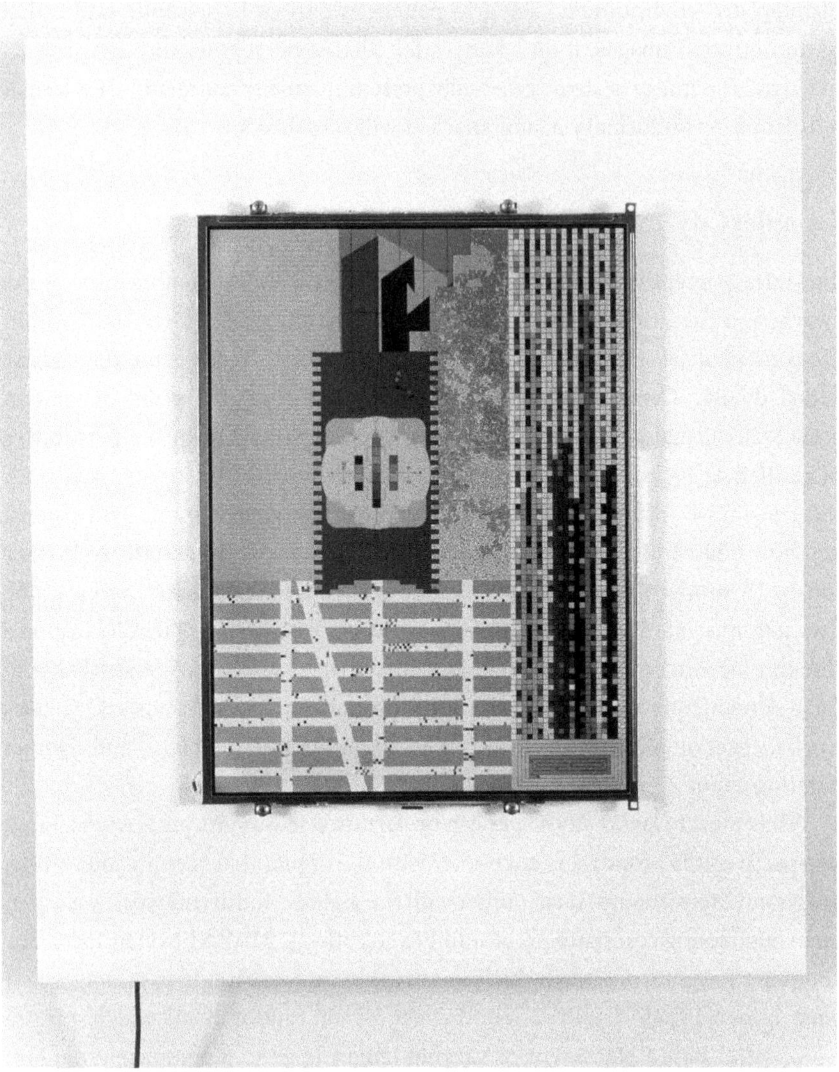

Figure 12 John F. Simon Jr., *ComplexCity*, 2000, Software, Macintosh G4 PowerBook, and plastic acrylic, 15.5 × 17.5 × 3 inches, runs continuously, never repeats. Courtesy of the artist.

look at Mondrian as a painter of systematic grids, but we have to be careful. The object-computer can work backward and become an illustration for a posthuman, detached way—faster than the speed of perception—of deciphering patterns in the body of work of a particularly apt painter.

However, Yve-Alain Bois offers this word of warning: "It goes without saying that this picture—like the classical neoplastic paintings in general—does not come under the heading of systemic or programmed art. But if it is not systemic, isn't it, in some way, systematic?"[74]—systematic in the sense that a method ensures a constant evolution of the grid that Mondrian is producing. Bois does, however, mention that Mondrian "loathed any axiomatization of his art" and consistently insisted that he worked by intuition and not calculation. Bois takes issue with an art historian who, while analyzing a Mondrian painting in a remark in a footnote, stated "that one of the strips is narrower than the others (the second vertical yellow strip form the right)."[75] Concluding that a strip looks "accidentally narrow" triggers Bois to unleash some vitriol about the author's ignorance of Mondrian's "microscopic precision."[76] But it does make our point about scanning the grid for inconsistency that Simondon reads through his concept of the intra-perceptive image. This is relevant, not because of the systematic versus the systemic, man versus machine, or the accidental line versus the deliberate line oppositions, but because, what is at stake is the plurality, the multiplicity, the volume in which the intra-perceptive image operates, in which algorithms sort through data that is too voluminous for humans to process. Such discussion of the volume and plurality of images leads us to look at algorithmic art in a new way when dealing with the reality of images that multiply in such a fashion. Sorting through them results in the very reconsideration of the nature of the image and the necessity of new ways of metaphorically understanding our relationship to them. Mondrian writes about the "the plurality of varied and similar forms" that do not "show contrasts" but where "syncopated" repetition can free us from objective forms (a priori shapes) and the way our subjective vision receives them (limiting in its own way). Bois explains that "to liberate our vision is also to accept that we no longer master it."[77] We are at the crux of the notion of intra-perceptive image which, according to Martin, is a visuality that does not belong to us but to the things in the environment.

Getting back to the grid and the abstract painter, we can turn, for the formal implication of the gridiron plan, to Peter Marcuse, who distinguishes between closed and opened gridiron. A closed gridiron plan is a complete and encompassing plan for a physically defined and bounded area; the open gridiron is an initial step toward plotting an unknown and perhaps unlimited area capable of indefinite expansion.[78] As for the influence of the grid in abstract art, we can clearly see the influence of Mondrian in Simon's work: Mondrian's *De Stijl*, in its utopian scope, was supposed to start with painting and spread throughout all art and design. The grid, the most harmonious pattern Mondrian could think of, was inwardly self-referential to the frame and flatness of the canvas while at the same time expansive, outward projecting, suggesting an infinity. Marcuse's contrast between open and closed grids is not unlike Rosalind Krauss's distinction between the centrifugal (outward-oriented) and centripetal (inward-oriented) grids of modern art that she applies to Mondrian.[79] These two functions of the grid described by Krauss are what provide the flux-image with the underlying structure of abstract art. But there is more.

Krauss, in her important essay *Grids*, takes issue with the avant-garde as being silent. It does not offer the viewer anything to talk about and lacks personal contact, mutely staring back, like Martin's clouds. Her essay asks why the grid pattern, so exclusively abstract and self-reflexive, became such a pervasive pattern in the art of the twentieth century. She specifically has in mind Mondrian's stubbornly repetitive grids. Her discovery in the essay is akin to what Marks finds in the virtual/actual duality of computer art: the grid pattern is but a skin that reveals, upon scrutiny, a hidden complex set of scientifically determined informational grids originating in nineteenth-century treatises on optics that inspired countless modernist artists of the twentieth century. In Krauss's discovery of the point of origin of the grid, we can see clearly a virtual element that has actualized into the patterns of abstract art—much like Hercules from the folds of the marble. Grids—those seen in treatises on optics studied by artists and only absorbed by habit, or habitude, as Leibniz would say—become the very subject of artistic abstraction, actualized images, with no obvious reference to their source. Krauss writes:

> An interesting feature of treatises written on physiological optics is that they were illustrated with grids. Because it was a matter of demonstrating the interaction of specific particles throughout a continuous field, that field was analyzed into the modular and repetitive structure of the grid. So for the artist who wished to enlarge his understanding of vision in the direction of science, the grid was there as a matrix of knowledge by its very abstraction, the grid conveyed one of the basic laws of knowledge—the separation of the perceptual screen from that of the "real" world. Given all of this, it is not surprising that the grid—as an emblem of the infrastructure of vision—should become an increasingly insistent and visible feature of neo-impressionist painting.[80]

The grid pattern, inchoate but pragmatic in optics manuals in the nineteenth century, migrates toward the surface of the canvas and soon becomes actualized as the only subject. This brings us back to the screen and Deleuze's view of information. This is the very grid of information that Deleuze borrowed from Steinberg and described as information-based art through the aesthetic of the folds. He writes that when art moves from natural representation to cultural abstraction, the canvas "becomes an opaque grid of information on which the ciphered line is written. The painting-window is replaced by tabulation, the grid on which lines, numbers, and changing characters are instructed (the objectile)."[81] We are firmly relocated onto the map, the city-information table, to be read from above, and not the window-countryside we look through.

As Deleuze demonstrates through his grid-screen pronouncement in *Cinema 2*, he was clearly inspired by Steinberg—who built his concept of the flatbed picture plane that holds information in response to the abstract emotional vertical spaces advocated by Greenberg. The cultural screen was to replace the natural window. But here, in relation to Mondrian, it is Greenberg who saw the shift in orientation coming. In fact, Greenberg writes that Mondrian's pictures "are no longer windows in the wall, but islands radiating clarity, harmony and grandeur—passion mastered and cooled, a difficult struggle resolved, unity imposed on diversity. Space outside them is transformed by their presence."[82] The window, vertical, is a horizontal island, a geographical image, radiating outward, centrifugal. Bois located this division already in Benjamin, who saw drawing as horizontal and painting vertical.[83] And, of course, Mondrian's New York suite is seen as a map from above since it marries itself perfectly to the gridiron, which leads Bois to call it a diagram, a plan, against human-centered perception.[84]

The modernist grid is here the opaque grid of information, that which signaled an organization of knowledge and that, in turn, became the subject of art. It is the algorithm behind the image, the virtual before the recognizable actual. The grid functions as a structure, a structure that Evens located already in Deleuze's virtual; and even if mute, as Krauss qualifies it, this soundlessness is in keeping with the unrepresentable undercurrent of virtual ideas that make up the matrix of the information age.

Conclusion

To conclude, I would like to briefly look at John F. Simon Jr.'s *Every Icon* (1997) [*Figure 13*] as an example of the elements that comprise the concept of the intra-perceptive aesthetics. Here is how the artist describes it:

> The piece consists of a 32 × 32 square grid where every square can be colored black or white. Every Icon starts with an image where every square is white and progresses through combinations of black and white squares until every square is black. The piece will show every possible image. Although it takes only 1.36 years to display all of the variations along the first line, it takes an exponentially longer 5.85 billion years to complete the second line.

A black-and-white grid made for the Internet, *Every Icon* is dependent on its electronic environment. It is a sublime work of art in its temporal reach and, like the progressive sketches within the intra-perceptive image, unfolds in time. It is geared toward a posthuman spectatorship, not only because humans will not be alive to see it when the first icon appears but also because it flickers too fast for the human eye. This piece relies on an algorithm, its articulation in our imagination, but this speaks to the two different ways that Simondon spoke of the virtual: an incomplete image but also one that triggers speculation. It operates through code and trigger. Once an icon appears, we will recognize it (if we are still there). And it is obviously a grid. The impact of this piece relies on the astronomical number of years it would take to fully appreciate. Time, on an imaginable scale, is what creates a sublime feeling in response to this work—perhaps baroque, but in a good way.

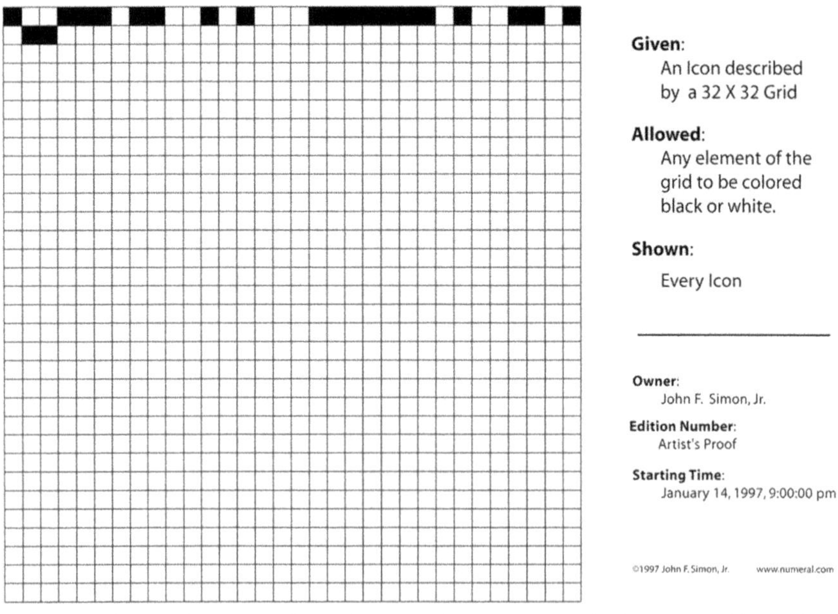

Figure 13 John F. Simon Jr., *Every Icon*, 1997, online artwork. Courtesy of the artist.

Notes

1 Marks, "Infinity and Accident," 38.
2 Laura U. Marks, *Enfoldment and Infinity: An Islamic Genealogy of New Media Art* (Cambridge: The MIT Press, 2010), 5.
3 Ibid., 5.
4 Deleuze, *The Fold*, 23.
5 Marks, *Enfoldment and Infinity*, 5.
6 James Williams, *Gilles Deleuze's Difference and Repetition: A Critical Introduction and Guide* (Edinburgh: Edinburgh University Press, 2013), 8–9.
7 Ibid.
8 John Mullarkey, *Post-Continental Philosophy: An Outline* (London: Continuum Press, 2006), 24.
9 Aden L. Evens, "Actual/Virtual," in *Understanding Deleuze, Understanding Modernism*, ed. Paul Ardoin et al. (New York: Bloomsbury, 2014), 247.
10 Ibid., 248.
11 Ibid.
12 Ibid.
13 Edward Willatt, *Kant, Deleuze and Architectonics* (London: Continuum, 2010), 148–49.
14 Gilles Deleuze, "The Actual and the Virtual," in *Dialogues II*, trans. Eliot Ross Albert (New York: Columbia University Press, 2002), 149.
15 Catherine Craft, *An Audience of Artists: Dada, Neo-Dada, and the Emergence of Abstract Expressionism* (Chicago: University of Chicago Press, 2012), 191.
16 Evens, "Actual/Virtual," 248.
17 Deleuze, "Actual and the Virtual," 148.
18 Deleuze, *Cinema 2*, 70.
19 Deleuze, "Actual and the Virtual," 150.
20 The temporal aporias of the crystal make the virtual and the actual visible at the same time: "Becoming, as lack of the present, is already at work in the first time-image; the present depends only on the crystal, it is an orginary secession of time that flies in and by the time-image." Marie-Claire Ropars-Wuilleumier, "Image or Time? The Thought of the Outside in *The Time-Image* (Deleuze and Blanchot)," in *Afterimages of Gilles Deleuze's Film Philosophy*, ed. D. N. Rodowick (Minneapolis: University of Minnesota Press, 2010), 23.
21 "About This Artwork: Rodney Graham, Torqued Chandelier Release, 2005," Art Institute of Chicago website, available online: http://www.artic.edu/aic/collections/artwork/185060

22 Christina Bagatavicius, cited on http://www.artic.edu/aic/collections/artwork/185060
23 Deleuze, *The Fold*, 9.
24 Christian Kerslake, *Deleuze and the Unconscious* (London: Continuum, 2007), 26.
25 Christine Buci-Glucksmann, "Temps et modèles du virtuel," in *L'art à l'époque du virtuel*, ed. C. Buci-Glucksmann (Paris: L'Harmattan, 2003), 10–11.
26 Walter Benjamin, "The Work of Art in the Age of Mechanical Reproduction," in *Illuminations: Essays and Reflections*, ed. Hannah Arendt, trans. Harry Zohn (New York: Schocken Books, 1969), 217–52.
27 Deleuze in Marks, *Enfoldment*, 3.
28 Ibid.
29 Deleuze, *The Fold*, 27, 36.
30 Hans Bertens, *Literary Theory: The Basics* (London: Routledge, 2014), 215.
31 Alexander R. Galloway, "*StarCraft*, or, Balance," *Grey Room* 28 (2007): 103.
32 John, F. Simon, Jr., "Swarms, 2002," John F. Simon Jr. website, available online: http://www.numeral.com/panels/swarms.html
33 Ibid.
34 Xin-She Yang and Xingshi He, "Swarm Intelligence and Evolutionary Computation: Overview and Analysis," in *Recent Advances in Swarm Intelligence and Evolutionary Computation*, ed. X.-S. Yang (Cham: Springer, 2015).
35 Ibid., 4.
36 Christian Jacob, Gerald Hushlak, Jeffrey E. Boyd, Paul Nuytten, Maxwell Sayles, and Marcin Pilat, "'SwarmArt:' Interactive Art from Swarm Intelligence," *Leonardo* 40, no. 3 (2007): 250.
37 John, F. Simon, Jr., "aLife, 2003," John F. Simon Jr. website: http://www.numeral.com/panels/alife.html
38 Rita Raley, "Statistical Material: Globalization and the Digital Art of John Klima," *The New Centennial Review* 3, no. 2 (2003): 74.
39 Ibid.
40 Pierre Lévy, *Qu'est-ce que le virtuel?* (Paris: Éditions La Découverte, 1995), 49, my translation.
41 Jean-Clet Martin, "Le virtuel incarné," in *L'art à l'époque du virtuel*, ed. C. Buci-Glucksmann (Paris: L'Harmattan, 2003), 208.
42 Ibid.
43 Ibid.
44 Ibid.
45 Mark Tribe and Reena Jana, *Art des nouveaux média*, ed. Uta Groseniak (Köln: Taschen, 2009), 86.
46 Ibid.
47 Ibid.

48 Gilbert Simondon, *Imagination et invention (1965–66)* (Paris: Éditions de la transparence), 120, my translations in all subsequent cases.
49 Jean-Hugues Barthélémy, "Glossary: Fifty Key Terms in the Works of Simondon," in *Gilbert Simondon: Being and Technology*, ed. Arne De Beover et al., trans. Arne De Boever (Edinburgh: Edinburgh University Press, 2012), 121–23.
50 Simondon, *Imagination*, 68.
51 Ibid.
52 Ibid., 72.
53 Ibid., 75.
54 Ibid., 76.
55 Ibid.
56 Ibid., 77.
57 Ibid., 78.
58 Ibid., 79.
59 Ibid.
60 Ibid., 66.
61 Ibid.
62 Ibid., 89.
63 Ibid.
64 Ibid., 91.
65 Ibid.
66 Ibid., 90.
67 Ibid.
68 Ibid.
69 John, F. Simon, Jr., "ComplexCity, 2002," John F. Simon Jr. website: http://numeral.com/panels/complexcity.html
70 Ibid.
71 Loe Feijis, "Divisions of the Plane by Computer: Another Way of Looking at Mondrian's Nonfigurative Compositions," *Leonardo* 37, no. 3 (2004): 217.
72 Ibid.
73 Ibid., 220.
74 Yve-Alain Bois and Amy Reiter-McIntosh, "Piet Mondrian, 'New York City,'" *Critical Inquiry* 14, no. 2 (1988): 252.
75 Ibid.
76 Ibid.
77 Ibid., 250.
78 Reuben S. Rose-Redwood, "Genealogies of the Grid: Revisiting Stanislawski's Search for the Origin of the Grid-Pattern Town," *Geographical Review* 98, no. 1 (2008): 52.
79 Ibid.

80 Krauss, "Grids," 57.
81 Deleuze, *The Fold*, 27.
82 Bois and Reiter-McIntosh, "Piet Mondrian," 258.
83 Ibid., 273.
84 Ibid., 275.

5

The Island/Image Apparatus: Virtual Networks in Kerbel, Bartholl, and Scott

The image of the island is emblematic of the concept of Utopia. Can the island also be a map-image applied to contemporary visual art, a tool of theory and criticism to explore the aesthetics of virtuality? And if it can be used as a critical tool, is it only in a utopian capacity? I will explore the philosophical concept of the desert island advanced by Deleuze. Perhaps it can serve as a guide to a set of specific spaces: insular space, utopian imaginary space, and space as capsule. First, I will analyze Janice Kerbel's *Welcome to Bird Island* (2001): a virtual island diagrammatically suspended on a web page that I will read through Deleuze and Peter Sloterdijk. Second, I will focus on Aram Bartholl's *Dust* (2011): a project to actualize a video game map into a solid three-dimensional structure in which notions of isolation and interface are filtered through the theory of Deleuze's and Paul Virilio's screens. Lastly, I will explore the function of cartography in Ridley Scott's dystopian *Prometheus* (2012) through the concept of mapping in Deleuze and Tom Conley. *Bird-Island* is an ensemble of documents. *Dust* is a proposal. *Prometheus* is a cartographic film. What do these three examples of maps have in common? They illustrate various aspects of Deleuze's notion of the island while offering clues about how to deal with the aesthetic of virtual space in contemporary culture. These works are part of an apparatus that connects and reconnects with its various segments to generate a fluid view of the map-image. In order to connect these heterogeneous works of art, I suggest that the island image must be studied in relation to the map, a concept crucial to Deleuze's philosophy. In that way we can use this island map for the purposes of delimiting virtual spaces. It will be an island-map assemblage, a *dispositif*, in Giorgio Agamben's sense of the word.

The image of the island Deleuze offers in "Causes and Reasons for Desert Islands" is certainly enchanting. He revisits the image of the island several times when discussing a model of thought based on an archipelago in *What Is Philosophy?*[1] or in *Logic of Sense*, where he focuses his attention on Michel Tournier's *Friday*.[2] The island image functions somewhat like another geographical image qua epistemological concept of space: the desert. The desert is equally as resonant as the island, and re-emerges as part of a geography of portraiture in *Logic of Sensation*[3] and in "The Shame and the Glory: T.E. Lawrence," where it is the setting of abstract ideas and "projected images."[4] But of course the desert has nothing to do with the desert island except as an instance of Deleuze's emphasis on liminal and extreme spatial conditions for models of thought. And perhaps the desert as a matrix for thought and primary condition for images can inform, in retrospect, the creative condition of the desert island. Even though the desert part of desert island is erroneously taken literarily (a desert island is not a desert but deserted[5]), the desert nevertheless seems to conjure elemental primary conditions for thinking at the end of Deleuze's corpus similar to those of the desert island at the beginning of his writing life.

And so, starting with these terms, "desert island," I will first, in guise of an introduction to this duality, explore how Deleuze's "Causes and Reasons for Desert Islands" foreshadows an aesthetic theory of space; the dualities that are at the core of the concept; the castaway as the figure necessary to establish the "desert" conditions of the island as well as the notion of "re-creation"; how re-creation, the foundational redundancy of the desert island, is entwined with presence and therefore establishes the notion of the virtual; how, according to Gregory Flaxman, human presence grounds the potentiality of the island; and finally, how, for Venessa Brito, potentiality is undeniably linked to artistic activity.

The imagery of "Causes and Reasons for Desert Islands" sets the stage for the importance of geography and mapping for Deleuze's philosophy.[6] And even though it could be argued, as Conley does, that certain key concepts of Deleuze's extended corpus are present in this early text in germinal form, they seem, at first sight, to resist any direct application to art and aesthetics. Conley, who surveys in detail the topography of Deleuze's "Desert Island," sees it as "a parable for the longer work on difference and repetition or a threshold, following his comparisons of eggs to Bodies without Organs, for the distinction of 'smooth' and 'striated' spaces."[7] For Conley, Deleuze's desert island is a metric for space theory: he sees Deleuze as a geographer and philosopher of space, reading the "inherent cartography" of the essay only insofar as he is concerned with space and imagination.[8] Flaxman echoes Conley by reading the desert island as proto-Deleuzean spatial theory; he sees a direct application of the essay to Deleuze's broader philosophy, since the essay bears witness to "a kind of geographic imagination that in some sense describes Deleuze's own philosophical development."[9] The island as image is already, at least turned toward Deleuze's corpus, indicative of a spatial theory. This spatial theory is based on a series of dualities that make up the image of the island.

Deleuze explains at the outset of an essay brimming with dualities that geographers contend there are two types of islands. The fact that there are two types confirms something already sensed by imagination. Geography as science renders imagination as the material for mythology, whereas the latter pair (imagination as myth) animates the former. These two types of islands are continental and oceanic; one drifts away from land mass while the other appears in the middle of the ocean, the result of volcanic activity. Their genesis awakens the imagination with strong imagery.

Another visual duality determines the island: earth and water. The relationship between earth and water, their hostile relationship, is elemental. This elemental quality is going to be eroded, according to Alphonso Lingis, in Deleuze's subsequent texts following his take on Tournier's reinvention of Daniel Defoe's *Robinson Crusoe*.[10] At this early stage of his corpus, Deleuze tells us that an island seems philosophically deserted, as the human being, in order to inhabit an island, must believe that the battles between ocean and earth have come to an end. The elemental is tucked away to carry on. Islands, then, are only "theoretically" deserted.[11] What the island represents, however, is the isolation of an originary point from which re-creation erupts.

But how to approach this human presence that responds to the elemental? At first sight, it seems that the human is the castaway crashing onto the shores of the island. Deleuze explains how a castaway seems to spoil the desert quality of the island. In fact, Deleuze rapidly corrects, the castaway does not disrupt the desert element of the island but makes it sacred. He seems to equate separation with creation, saying that a more separate and more creative human gives the island a "dynamic image of itself."[12] Paradoxically, the island, having been populated, becomes conscious of its desolation: "The island would be only the dream of humans, and humans, the pure consciousness of the island."[13] Geography and the imaginary become one in such a paradoxical "movement."[14] The inhabitant of a desert island is the virtual human, a self-reflexive human: "There you have a human being who precedes itself."[15] In this case, Defoe's desert island is cause for chagrin because "everything is taken from the ship. Nothing is invented."[16] The island is a stencil of an established society, not a map of things to come. The text announces prototypically the negative/positive charges of the tracing/mapping terminals Deleuze and Guattari intertwine into a circuit in *A Thousand Plateaus*.

The island is re-creation, not the originary creation but a redundancy. As Elizabeth Berkebile McManus explains, the issue of re-production is involved with time: "A re-beginning indicates that the island must build on the past, must engage what came before it in order for it to move towards the beginning anew."[17] What we are dealing with here is a landmark to presence—a series of moves through which the island's desert or the castaway's presence precedes itself or becomes a memorial to itself. But this *jeu d'esquive* of presence is grounded on vision, real or imaginary. It is a perverse presence, in Lingis's terms, since it is an existence lacking the "other" as a "master category": what is meant by the other in this case is not an actual individual who can share one's experience but rather the other as category—a virtuality: "As the system of possible perspectives maintained about the actual, transforms every pattern from a plenum that mesmerizes the solitary gaze into an objective."[18] This virtual is given an image: a space completely filled, plenum. The kaleidoscopic patterns it yields are the focus of the theoretical castaway's view. They are like the honeycomb light patterns in a film as the camera crosses paths with a sunbeam—themselves contemplated rather than the object they shield. We could call this a lens flare within experience. Or, as in Deleuze's later view of the desert, the virtual is the haze from which perception emerges.[19]

But the castaway, whose presence puts into motion a series of relationships between creation and vision, also illustrates more fundamental relations between the human subject and the geographical ground. Flaxman explains how the idea of the island is motivated by human interests. Islands are located "in the leaving behind of one world or the founding of another."[20] But these human motives set aside "the geological and geographical movements of the earth," because, in order to inhabit the island, the human must believe the elemental battles representing the island have subsided. This ignorance of what the island represents shifts the relationship between subject and ground: "In point of fact, all islands are deserted or, inversely, only on the condition of its desertion is an island really an island, and this proposition immediately deterritorializes ground and subject alike."[21] Flaxman reads in this image of the island a finitude, and end point to our presence, which he calls a *memento mori*. The ground of the island, its geological and geographical ground, is what becomes its orientating matrix. But in the absence of the human, there is a potentiality of a people to come.[22] And so, as Flaxman suggests, instead of "grafting civilization" onto the desert island, Deleuze wants us to imagine a people in "mutual resonance" with the earth.[23] Geography and imagination have the potential to come together because of this resonance ("sufficiently separate, sufficiently creative") between people and the island.[24] Is this a utopian vision in Deleuze's philosophical program?

Tightening the shot on aesthetics, another interpretation of human potential can be read into the presence on the island yielding an artistic creation. Venessa Brito makes a convincing case for the relationship between Deleuze's aesthetics, the image of the island, and the molecular space of art. For Brito, the kaleidoscopic patterns of the haze out of which perception emerges have a human shape. Writing about the desert island, Brito says that the only possibility we can consider is of a "molecular people, a phosphorescent people made up of dancing grains and luminous dust that was already there at the beginning of the work of art."[25] Like Flaxman and Berkebile McMamus, Brito sees this inchoate, virtual people already on a seemingly deserted island: "this atomic people of the desert island whose dance is confused with the mythological combat between elements of nature."[26] But could the island function more pragmatically? The patterns are people: pure virtuality. The combat between elements of virtual patterns of aesthetic potential are confounded. In contrast to Flaxman's close reading of Deleuze, in which he finds a deterritorialization of "ground and subject alike,"[27] Brito reads the people as already part of the island but in a state of pollen-like molecularization. Yes, the human is already grounded on the island. But what kind of human? Brito's image of being as milieu can be read through Simondon's pre-perceptive stages of the coming into being of the image. We are in a spot similar to Deleuze's desert, where the hazy state provides an environment from which perception is born, itself at the stage of rough outlines.[28]

For Brito, the island is a plane open to artistic potentialities. The phosphorescent people, atomic grains registering luminescence as a trace of the past, recall Deleuze's own description of the virtual. She reads the island as a matrix of art. A stage full of utopic potential that functions within art. An island in a state of "creative condition."[29] In fact, this creative condition is at the core of Deleuze's own philosophical mission: "The aim of philosophy is not to rediscover the eternal or the universal, but to find the singular conditions under which something new is produced … [It] aims not at stating the conditions of knowledge qua representation, but at finding and fostering the conditions of creative production."[30] This condition for creativity is the desert island: "Because we can imagine it as separate from the rest of the world, an island, Deleuze suggests, is a place from which the world itself might be created anew."[31] The island, with its paradoxical shell-game of presence/absence, engages in a mechanism of creation determined by a geographical location, real or imaginary.

But why the island image? This image offers a way of expanding on the notion of the frame so central to art; but with dimension and dynamism to properly capture art, it situates itself historically beyond van Gogh's shoes attached to the earth and to the world as framed by Heidegger to reveal the function of art. Conley reads the island as "selected, classified … isolated or pigeonholed."[32] The re-creation is prompted by the isolation: the island stands in for repetition and variation. The island is not created (continental, oceanic); but rather the mechanism of its framing is what causes it to be discovered, or unconcealed: "The island is found within the process that makes it liable to be contemplated in the first place."[33] This idea of frame and isolation is one that will be clearly expounded by Sloterdijk. The island image is a marriage of the significant place of the map in Deleuze's thought and a way of rethinking the frame so vital to art. As we will see, these elements will become crucial to the following analysis of *Bird Island.*

Welcome to Bird Island

The website www.bird-island.com opens on a splash page containing a small centered photograph of the aerial view of a crescent-shaped green island in the middle of the blue water. The white page around the photograph has the effect of an expansive nothingness surrounding the frame, the way that the sea surrounds the island. The words "Welcome to Bird Island … " written below the photograph underline the white "page" effect of the neutral background while simultaneously, through the ellipses, indicate something with an expansive spatial potential. Click on the page and you are treated to a static, documentary website. The text situates the island "'Just above the tropic of Cancer' in the Great Bahamas Bank" [*Figure 14*]. It tells us the island is unspoiled. The bar next to the main text displays "welcome," "the island," "the exumas," "the villas," "the investment," and "registration" buttons [*Figure 15*]. A list under the "birdlife" tab under the "island" opens to list and categorize—according to c-common; fc-fairly common; r-rare—the birds of the island. Another tab

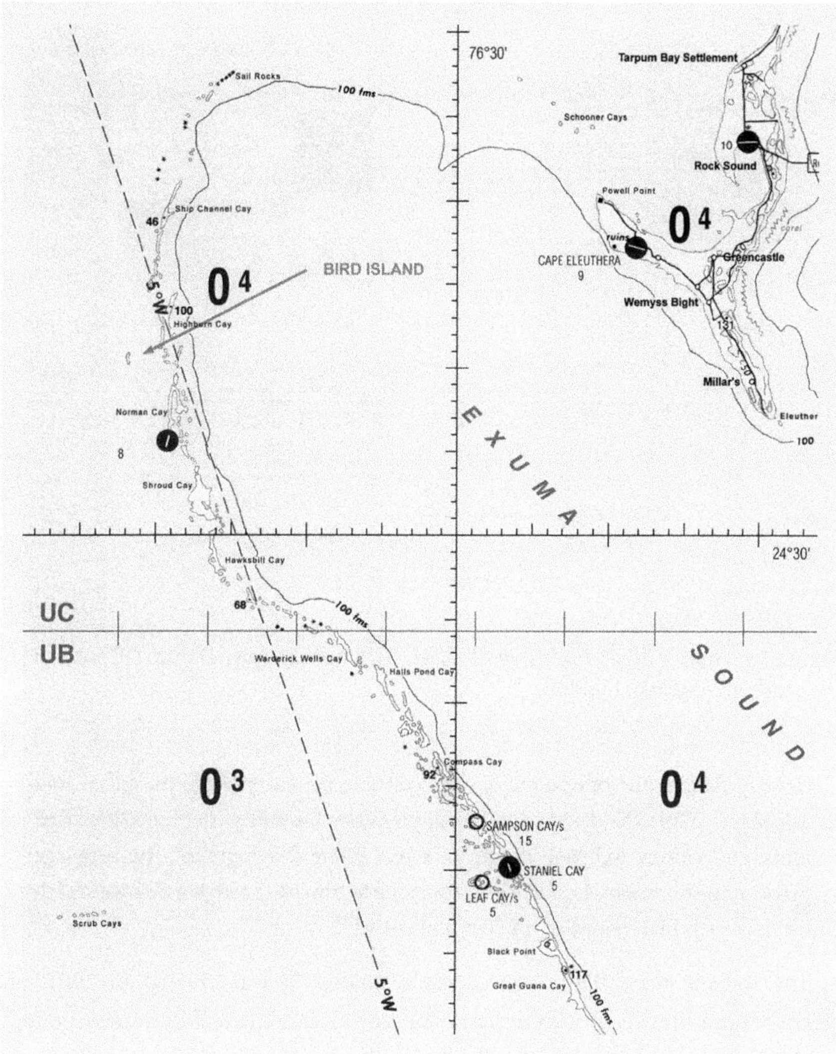

Figure 14 Janice Kerbel, *Bird Island Project*, 2000 location map (digital print), Digital print: 26 × 32 cm (10 ¼″ × 12 ⅝″). Courtesy of the artist; greengrassi, London.

shows its beaches. You can click and gather all of the information about this island and its real estate opportunities, provided to you in an enthusiastic yet deliberate text.

The website is functional and concise; the pictures have a stock-photo quality. This is, of course, on par with the commercial feel the artist aimed for in this work, which

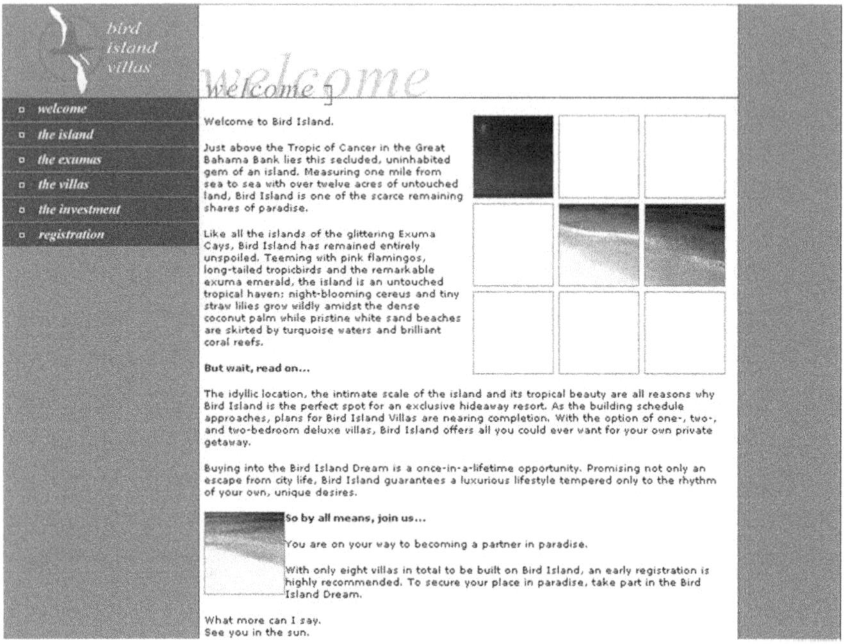

Figure 15 Janice Kerbel, *Bird Island Project*, 2000 website (digital print). Courtesy of the artist; greengrassi, London.

> cleverly mimics the online marketing strategies of real estate time-shares sold "off-plan." While the design and layout pastiches the soft pastel hues, italicised fonts, and clunky webpage design of a real estate development, the language plays on the boredom that may lead someone to the site through a Google search for keywords like "paradise" or "tropical island."[34]

The website itself has a time-capsule quality. It was created in 2001, an eternity ago in terms of web design evolution. It has the feel of a message in a bottle from another era. Words like navigating and surfing come to mind as a tongue-in-cheek way of tautologically exploring a website-island. The site lists detailed information about history, geology, flora, and fauna. The whole thing is marketed as a desert island—but the villas (the architectural plans provided) offer the first clue that this island is not meant for one inhabitant. Perhaps it is only when an interested party fills out the registration form that the full irony of being the inhabitant of a "bird"—or "desert"—island is finally felt. *Bird Island* is an information-era meditation on the notion of isolation. In the following, isolation is explored through the insular impulse based on the fragment and the cut. We will see how isolation informs the concept of the frame, so important to art, and, in turn, how the frame introduces the idea of artificiality. The insular, frame, and artificiality result in the notion of control that determines a strange utopia at work in *Bird Island*. Kerbel's work conjures images that connect with the idea of control-based utopia (seen in other Utopian islands such as Thomas More's and Jules Verne's) which seal island aesthetics around the island of the museum which is significant for the aesthetics explored therein.

Isolation is one of the elements that we can apply to a website island: the encounter with Internet art is not necessarily a solitary experience. Some works of Internet art conserve the traces of the viewers to become asynchronous interactive experiences; but the case of *Bird Island* carries no such traces: we truly are the solitary inhabitant of the island coded by Kerbel.[35] The viewer is asked to conjure the possibility of the island in an art work disguised as a commercial pitch—but its visitors arrive at the end of the website only to realize that the desert island dream cannot be realized.[36] It is not a website of an island; rather, the island unlocks the insular potential of the virtual space of the Internet's imaginary layout.

The insular impulse at the heart of Kerbel's virtual island can be read through Rosalind Krauss's assessment of Alfred Steiglitz's "cut" as a self-reflexive marker of the photograph[37] and Maurice Blanchot's view of René Char's fragmentary poetic style as determining the relationship of words and the space on the page where the fragmentary word is wrenched away from a whole, like a meteor from the sky. It is striking that Blanchot's concept of the fragmentary is associated with the island image: "Speech as archipelago: cut up into the diversity of its islands and thus causing a surging of the great open sea; this ancient immensity, the unknown always still to come, designated for us only by the emergence of the earth's infinitely divided depths."[38] Here, the question is one of "re-creation" of an originary "immensity."[39] The oceanic immensity can be compared to the immensity of the sky in Stieglitz's photographs of clouds where the limit of the picture isolates, cuts, and frames.

Like Stieglitz's photographic framing of clouds, the island is the geographical equivalent of the artistic frame. Sloterdijk, in his *Sphere* theory, imbues the island with a framing power: "It is the framing power that draws a boundary to restrict the rising power of the island, as if these surfaces without context were some emergent natural works of art contained by the sea like showpieces of nature."[40] What the frame is doing, in fact, is "closing the work of art off against the surrounding world and holding it together. The frame proclaims that a world is located inside of it which is subject only to its own laws."[41] Sloterdijk, like

Lingis, Conley, and Brito, thus treats the concept of island as a creative template, but he does so by equivocating geographic isolation with artistic framing.

The frame underlines artificiality: "Whoever wishes to understand the island must build prosthetic islands that reproduce all essential characteristics of the nature island point for point in the technical replica."[42] "What would be the purpose of such an exercise?," asks Sloterdijk. Instead of a hopeful program of relocating the earth's population onto another planet, the possibility of a prosthetic planet reveals the unrealizable reconstruction of an original environment: "The repetition of life elsewhere shows how much of life was understood in its first manifestation."[43] This is an interesting and pragmatic take on Deleuze's re-recreation. In Sloterdijk's expanded definition of what constitutes an island, the island is artificial, like a space capsule or a greenhouse, and it is a restricted space where the management of the atmosphere conditions the living situation.

The artificiality is triggered by a control impulse manifested in the frame. The frame is not simply a window displaying an illusion, but a constraining mechanism: "What the frame does for the picture by excluding it from the world context … is carried out for the island by the isolator—the sea."[44] In fact, the island is this matrix for art because of its separation and re-enactment of the world: "If islands are world models, it is because they are sufficiently separate from the rest of the world context to harbor an experiment about setting up a totality in a limited format."[45] The sea isolates and creates a maquette of the world determined by its "insular climate,"[46] not just a boundary line, but an invisible yet crucial dimension. Climate becomes a critical factor as it reorients the island: retaining its framing, isolating function while discarding the limits of the frame/object: "The absolute island requires three-dimensional isolation—and thus the transition from the frame to the capsule, or, to use an analogy from art, from the panel painting to the spatial installation there can be no complete enclosure without vertical isolation."[47] The island as a model of aesthetics allows us hold up a new type of image, one that is not held up in a rectangular frame but instead determined by an internal spatial coherence. The island model is a hermetically sealed milieu with its own artificial atmosphere, recreating conditions of its own subsistence: an image-milieu.[48]

Utopia as controlled milieu

Utopic isolation of the island also depends on it being artificially framed and controlled. We see this in Elizabeth Grosz's utopia, suspended out of time and out of space, in the geographical arrangement of the island in Thomas More's *Utopia*, as well as in Jules Verne's *Île à hélice*, a science fiction novel that takes place on an island with propellers (as the subtitle announces, a boat island for billionaires).[49] It is Anna Lovatt who makes the direct connection between Kerbel's island gardens and the dystopia/utopia duality. In creating an imaginary island, Kerbel engages with one of the dominant tropes of utopian thought since More's foundational *Utopia* of 1516. Indeed, Kerbel's island takes a form similar to that described by More: a crescent shape curving in on itself, protecting a large sandy bay from the elements.[50]

To what extent does Kerbel's work figure within a critical utopian theory? Grosz has explored the issue of the utopic in relation to space: "The utopic is always conceived as a *space*, usually an enclosed and isolated space—the walled city, the isolated island, a political and agrarian self-contained organization, and thus a commonwealth."[51] Grosz focuses on the issue of control when considering the idea of the spatial utopic, specifically on how, as a commonwealth, it can be controlled within its walls: "This emphasis on place and space is no doubt why the utopic has been a locus of imagination and invention for architects, as well as for political theorists, activists, and fiction writers: descriptions of buildings and municipal arrangements figure quite prominently in Plato's, Aristotle's, and More's accounts of ideal political regulations."[52] What Grosz proposes instead is another reading of the utopic based on time and the virtual: "the utopic as a dimension of the virtual, and admixture of the latency of the past and the indeterminacy of the future, the mode of linkage between an inert past, conceived as potential, and a future not yet in existence."[53] But the notion of a sequential time is problematic too. The utopic, since it has no telos, no possibility of a future, instead relies on the control of the event: "While a picture of the future, the utopic is fundamentally that which *has no future*, that place whose organization is so controlled that the future ceases to be the most pressing concern."[54] Is Kerbel's *Bird Island* utopic because controlled: isolated, researched, delimited? Or because of its self-defeating utopianism, an anti-utopianism that resides in the very notion of utopia—Kerbel's island has no future. When you set foot on it, it ceases to be itself in its advertised essence. Past and future, place and non-place, combine to form a virtual vision of the utopic.

From a geographical perspective, it seems that More's island of Utopia is already set up to control unruly weather. More writes about the crescent shape of the island, remarking how the middle breadth and narrow tips of the crescent "do fashion the whole island like to the new moon."[55] The island's shape, as More explains, keeps the weather calm and the sea smooth, which comes with its financial benefits:

> Between these two corners the sea runneth in, dividing them asunder by the distance of eleven miles or thereabouts, and there surmounteth into a large and wide sea, which by reason that the land on every syde compasseth it about and sheltereth it from the winds, is not rough nor mounteth not whith great waves, but almost floweth quietly, not much unlike a great standing pool, and maketh wellnigh all the space within the belly of the land in manner of haven, and, to the great commodity of the inhabitants, receiveth in ships twoads every part of the land.[56]

We have already compared the island's shape—the financial element at the core of utopia—to that of Kerbel's *Bird Island*. What makes this comparison even more striking is the relationship between the elements of the cities, which have a regularity only found in Kerbel's controlled environment. Another way of linking the artificial island with control is through Jules Verne's *L'île à hélice*; in fact, one feature of the island that underscores its artificiality is its ordering of nature in a grid pattern, like a chessboard of sorts:[57] "The city was built on a regular plan. The avenues and roads, provided with verandahs above the pavements, crossed each other at right angles."[58] One is reminded of Hubert Damisch's assertion of the potentiality of the chessboard pattern: "A grid, a chessboard, in this case, is not a structure, but the possibility of it."[59] Because the grid, as Damisch explains, since it first appeared to order intellectual data, is the starting point of historical becomings—in other words, it is not a structure but a diagrammatic approximation open to a potentiality. Furthermore, the grid pattern, for Damisch, is "emblematic of the nostalgia for utopia."[60] Nostalgia for utopia is a nostalgia for "delimited space in which the game of history would take place and make sense."[61] In fact, the notion of control is part and parcel of the potentiality of the grid: "A space one could be in control of, a game one would know how to play."[62] A potentiality in keeping with the never-attained utopic goals of artificial islands. Apart from city planning, Verne's island, too, has a controlled artificial climate and environment: "The squares, placed at the intersections of the main arteries of the city, carpeted with lawns of English freshness, whose flowerbeds, where the essences of temperate and torrid zones intermingled, did not yet inhale from the soil enough vegetative force."[63] All is new artificially implanted surface, yet familiar, controlled.

Propeller island is the logical conclusion to the elements of Crusoe's mercantile bourgeois island, as Deleuze laments. But it also brings to the scene the notion of art. Of course, price is no object for these passengers of the island, which itself becomes a compendium of European art. Italian, Dutch, German, French art, Verne tells us, are featured in the museum; it contains paintings by Raphael, da Vinci, Giorgione, Correggio, and many more—Verne names over forty artists, stretching the list to about half a page. The island is a capsule of culture.[64] He does say, with some relief, that "the impressionists, the anguished, and the futurists" have not yet cluttered the museum.[65] But in keeping with the notion of capsule-island-climate, Verne states that the artworks are safe there because of the climate. Sculptures and statues are not affected by fog or rain. The island as isolator does not recreate its own atmosphere, but skirts bad weather by pulling anchor when it threatens. The outside atmosphere of this artificial island can be adjusted through its wanderings in different seas.[66] We are reminded of Deleuze and Guattari's statement about art and preservation: "Art preserves, and it is the only thing in the world that is preserved."[67] But here we move from the work of art in itself as preserved within the frame and the museum as preserving atmospheric environment.

How did we get here? We started with an island self-reflexively announcing itself. It was a space for people to come. It was a sight of creativity, of re-creation. Then we saw an artist, Kerbel, who captured these issues in a paradox: the island is out of time, a spatial utopia behind the screen. Step on it and you ruin the beach; it is not a desert island anymore. The image of the island works through framing and isolation, but it also presents notions of the beginning of art—the elemental conditions of art. Art, to reiterate, is not simply spatial: art is a milieu. And this milieu is created by isolation, be it in the form of a website or a museum. Isolation also threads through to the next artwork analyzed: from virtual, online sources to material display in a (potential) museum, an island that stands between the ephemeral and the monumental.

Dust

I will now shift from the ocean's shore at the edge of our desert island to look at a punishing desert. Aram Bartholl's *Dust* (2011) is foreboding [*Figure 16*]. The work's point of origin is a game map from the video game *Counter-Strike*, which takes place in a sub-Saharan combat theater where terrorists and counter-terrorists battle. The art work is a plan for a concrete sculpture of the virtual space in which the video game battle takes place: concrete as in the

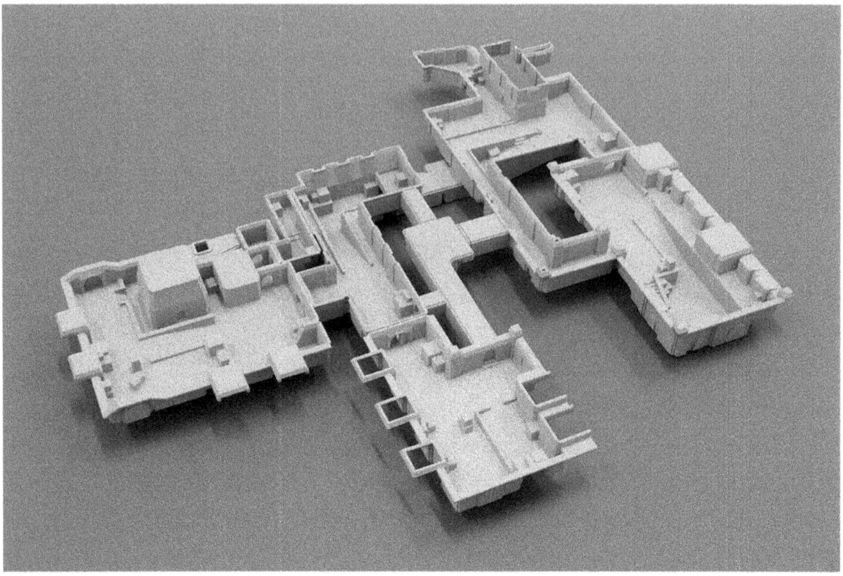

Figure 16 Aram Bartholl, *Dust* (2011). Sculpture, Alumide 3D-print, 36 × 33 × 4,5 cm. Courtesy of the artist.

actualization of a virtual space, but also made of concrete, the construction material. It should be mentioned that the piece does not exist yet. *Dust* is a plan for a massive, human-scale architectural sculpture. But it is not a shelter: its model is a more or less open space, of which the end product will be an imprint. It is a concrete construction associated with warfare. Its austere aesthetics also bring to mind a bunker. The work is a massive blind and dumb mold of negative space—of a place that does not exist materially—extracted, pulled, from behind the screen. Of course, this work perfectly illustrates the notion of the map-image: virtual, potential, real yet immaterial, spatial insofar as it is mediated by a screen.

It does, however, share some characteristics with Kerbel's desert island. It is finite—an isolated space without links to a larger environment—and therefore appears to have the characteristics of an isolated island. But this island is the fossil of a virtual space rather than an actual space. Like Kerbel's Internet art piece, *Dust* also has a critical relationship to the digital, in this case a video game. But the final work is not yet made. It currently exists as a plan with several maquettes of paper and some test samples made out of concrete (a crate that appears in the game has been cast in concrete to give it the air of a minimalist work), and, in 2013, Bartholl recreated part of the game space of *Counter-Strike* in paper on a human scale. In plan form, it complicates this work's relationship of the abstract and the concrete and of the virtual and the actual work. Both Kerbel and Bartholl engage with the notion of habitability: If Kerbel wanted to give the initial illusion of a habitable space, Bartholl's space is to make a virtual space habitable.

All of these elements are part of Bartholl's complex and humorous take on the new realities dictated by the digital environment. He is an intriguing artist who simultaneously celebrates and critiques our sitting on the shore of the actual while dipping our toes in the waters of the virtual. In another piece, entitled *Map* (2006–13), Bartholl makes enormous public sculptures in the shape of the graphic "pin" icon on Google Maps that points to a searchable place. Bartholl plays with the *trompe l'oeil* effect of shadowy pins on a flat map. In *15 Seconds of Fame* (2010), Bartholl chases a Google Maps car he sees passing by a café he is patronizing and becomes part of the street view of that particular location. Bartholl's work, then, is yet another exploration of our straddling of both digital and material realities. This zone between the digital and the material is examined through Heidegger's concept of the jug that helps us delimit a negative space and the concept of the virtual assessed through ballistics and optics by Galloway, Deleuze, and Virilio.

We have already invoked Heidegger in relation to an environmental analogy to art. Here, though, a spatial analogy is in order: Heidegger's discussion of the thing as a space in a jug. What is the thing here? It is the space of an Internet site, a virtual plan, an empty cavity whose outline has been rendered full and solidified. But not yet, since we are only at the planning stages. Bartholl's project questions the notions of space, virtual terrain, map, and plan. Since the content of the work of art is the map, the environment this map traces is going to be given a negative spatial and architectural treatment; but as this project is only in the planning stages, it is still virtual, still not actualized.

The relationship between the virtual space of the *de_Dust* map (map in *Counter-Strike*) and the *Dust* project (artwork plan by Bartholl) can be compared to that of the actualization of the void that Heidegger works through in his reflections on the thingly character of a jug in his essay *The Thing*. The jug is an interesting object insofar as it is a container with a function. The jug is not just a material object; it does something: "What is a jug? We say: a vessel, something of the kind that holds something else within it."[68] The jug reveals its nature when it is filled, Heidegger argues: "We become aware of the vessel's holding nature when we fill the jug."[69] Here, the purpose of Bartholl's work is revealed: to fill out the virtual space of the game map and to actualize it in concrete—in effect to make viewers aware of the relationship between a space that is nonmaterial and one that is. In fact, the more we think of the space of the game map as a virtual meeting space, the more apt Heidegger's jug seems as a metaphor for Bartholl's artistic gesture: "The emptiness, the void, is what does the vessel's holding. The empty space, this nothing of the jug, is what the jug is as the holding vessel."[70] Heidegger asks whether the potter shapes a jug and answers that, in effect, he does not: "No—he shapes the void."[71] Bartholl also shapes the void, the virtual void of *Counter-Strike*, to make us understand the relationship between virtual and actual space. Giving concrete appearance to the virtual void that is shaped by ones and zeroes, Bartholl seems to reveal, like Heidegger, something of the nature of the virtual game space, because, as Heidegger says, "We represented the effective features of the vessel, that which does its holding, the void, as a hollow filled with air … but it is not the jug's void."[72] How does it hold, this void that is not the jug's—or in Bartholl's case, the void that is not the sculpture's, since it is also a representation of the game space? The nature of the void here is different, one holding millions of users, while the other one will be limited to the number of bodies that can fill the receptacle of the walk-in sculpture. The negative space is the actual work. It is like Styrofoam from the box of an appliance, easily discarded because it is not the main event of the product's box, but that which has brought it safely to our doorstep. Here it is the gaming experience that is the main event, and not the empty space that contains it. Yet they are interdependent.

The concrete in Bartholl's case is also an evocative material; as far as possible from the virtual space behind the video game player's screen, it nevertheless seems to communicate something less solid than its own materiality. The concrete serves to solidify abstract game space into a museum object that commemorates the game space as a cultural object.[73] Take for example the work of another artist whose use of concrete yields to contemplation of less material concerns related to her subject matter: Rachel Whiteread's *House* (1993). If Bartholl's work does not yet exist, Whiteread's house does not exist any longer. Whereas Whiteread's *House* was a "memorial to memory," Bartholl's work will be a monument to virtuality.[74] We can compare these works on the level of the material: concrete. The negative/positive molding: the inside is turned outside. On Whiteread's *House*: "Its inviolability is exposed from the space being turned inside out, its living history entombed in concrete."[75] A sepulchral aspect comes through: it looks like a funeral monument for what was once lived in the house, not the house itself. In this way, Bartholl's *Dust* will also be a monument, not to the actual game or the desert city streets, but to the virtual experience of millions of users playing the game online.

Bartholl's project involves a concrete sculpture that would span more than 12,000 square meters of space. The artist explains that his work will be a "1:1 scale replica of one of the most played computer game maps in the world. The idea is to build the 3D model of 'de_dust' of the first person shooter game 'Counter Strike' as a permanent 'building' from concrete, making this map accessible as a large scale public sculpture."[76] The artist was drawn to the video game as the initial medium for his work because the games involve an initial dependence on space that other media do not: "Computer games differ from other mediums such as books, movies or TV, in that spatial cognition is a crucial aspect in computer games. To win a game the player needs to know the 3D game space very very well."[77] The virtual space of the video games draws in millions of users who experience a similar space together. That is why, Bartholl states, he does not treat the video games themselves but rather the space within them as the object: "Computer game architecture and game maps have become a new and yet undiscovered form of cultural heritage."[78] The space seems to be an artifact. The experience of the video game is more real to some than nonspace, such as the airports described as a variation on Foucault's view of indeterminate space: "A computer game map like 'de_dust' appears to be more real than many other places in the world such as artificially constructed places like supermarkets, airports or cities like Dubai."[79] A game map is contrasted to such nonspaces described by Marc Augé (non-site) and Michel Foucault (heterotopia). Augé, for example, writes of the elusive status of a space to be a site or non-site. He sees this relationship between spaces as a palimpsest.[80] For Augé, spaces such as airports feature on a list of non-sites that provide individuals with "another image" of themselves.[81] Foucault, in his essay "Of Other Spaces," provides a categorization of nonspaces such as heterotopias (spaces at the edge of established official, social sites) to qualify generic spaces, among them travel terminals, as sites of relations.[82] These spaces are then characterized by their fluid nature located in-between meaningful spaces. It is not without humor that Bartholl suggests a digital meeting point as being more concretely appreciated by its users than a real but generic place.

Bartholl picked a game from the 1990s since these have a built-in limit, unlike, as he explains, contemporary games, which have far more graphic power to generate limitless spaces: "Unlike current computer games (with their endless worlds and terrains), game spaces of the 1990s were still limited in size due to graphic card and processor power limitations. A respectively small and simple map like 'de_dust' offered a high density of team play with repetitive endless variations." The delimited territory of the game space creates an island, and that island, in turn, becomes a museum piece: "Made from concrete in 1:1 scale, the map becomes an art piece and a museum for a game at the same time. Visitors are invited to take a walk in materialized virtuality and experience the loaded game space in the physical reality."[83] Bartholl's piece therefore also raises interesting questions about the interface between the screen and the concrete, "the petrified moment of cultural game space heritage."[84] Furthermore, we are reminded here of Deleuze's hodology of the cartographic art, which replaces the monumental with the hodological.[85] With Bartholl's (potentially) massive piece, it is the monumentalization of the hodological that yields an archaeology-art for future commemoration.[86]

What the game space has in common with Kerbel's *Bird Island* is the screen as a mediating membrane that reconfigures the visual experience of the space: we do not read the information on the screen passively but are engaged with the visual elements actively. And the video game offers ways of looking through the screen as this digital space. Strangely, as we will see, some of the video game qualities of interacting with the gameplay have been intimated by Deleuze's creative ontology of the island. Galloway, whom we have already encountered in a previous chapter, elaborates on these active visual aesthetics in the digital virtual world by looking at the space of video games. Galloway analyzed the "ballistics and optics" in *Counter-Strike* as a space constructed around the rays that presumably beam out from the surface of the screen (usually represented by crosshairs or other such device): "In *Counter-Strike* the methods by which objects are selected and relationships are established between subjects and predicates flows from the simulation of linear 'rays' that are extended through mathematical matrices in such a way not dissimilar to the projection of a ray of light through space."[87] Galloway organizes the game space according to a starting point, the screen, which can be made synchronous with the eye if one extends the ray analogy, as he does, to the pre-modern theory of eyesight and light. The way that the game functions is similar, Galloway tells us, to "the discredited, pseudo-scientific notion that, in the faculty of human sight, rays extend from the eyes outward onto objects."[88] The rays emanating from a point on the screen hit their target. The game here is based on the trajectory of the eye and a bullet within the game space: "The targeting technique in *Counter-Strike* is what graphics programmers call 'picking,' a method of ray tracing in which a single ray is projected from the foreground, typically represented by a crosshair or other surrogate cursor, into the three-dimensional scene."[89] So, for Galloway, the game revolves around a very different interpretation of the optical than Bartholl's. One vision is of an extension of body through ballistics; the other is an experience of space through habit (rote).

Deleuze's analysis, in *Logic of Sense*, of Tournier's *Friday*, and specifically the action of vision that uncovers the Other in the field of vision, could be interpreted through a first-person shooter game aesthetic:

> I regard an object, then I divert my attention, letting it fall into the background. At the same time, there comes forth from the background a new object of my attention. If this new object does not injure me, if it does not collide with me with the violence of a projectile (as when one bumps against something unseen), it is because the first object had already at its disposal a complete margin where I had already felt the pre-existence of object yet to come, and of an entire field of virtualities and potentialities which I already know were capable of being actualized.[90]

Compared to Galloway's analysis of vision through the interplay of screen, this way of looking at objects has the deliberation of something painstakingly thought through in a game space. Furthermore, the idea of harm and injury, of something unseen attacking, has the hallmarks of a first-person shooter game. But what is key here is that Deleuze links his idea of the Other to the idea of virtuality and actuality.

Here, merging from the background are potentially harmful objects: injury, collision, the violence of general ballistics. We are not in the landscape of a vertical frame nor a map of a horizontal surface: "It is not enough to say that it is a landscape and that it lays out a place or territory. What it lays out are paths—it is itself a voyage."[91] It is a fluctuating map-image instead of a traditional view of the map. Another way of looking at this shifting representation of space is through the aforementioned hodology. Understood as the study of pathways, this psychological concept is applied by Deleuze to art (Deleuze has Carmen Perrin's environmental site-specific works in mind). The frame is bent into an island image, isolating, submerging, self-reflexive (not desert, if we are on it). This hodology applies to gameplay or the path one takes in a potential monument. The paths and the isolated space are necessary to create a virtual, digital, isolated but not frame-dependent work of art: be it *Bird Island*, where interaction and the paths the viewer takes through the document shape the work; or *Dust*, a potential, unformed conceptual piece that depends on a digital space. The virtualities of the distances rise from the page, the map, like three-dimensional "computer" models:

> A map of virtualities, drawn up by art, is superimposed onto the real map, whose distances [*parcours*] it transforms. Such internal paths or courses are implied not only in sculpture, but in any work of art, including music: in each case, the choice of a particular path can determine a variable position of the work in space. Every work is made up of a plurality of trajectories that coexist and are readable only on a map, and that change direction depending on the trajectories that are retained. These internalized trajectories are inseparable from becomings. *Trajectories and becomings*: art makes each of them present in the other, it renders their mutual presence perceptible.[92]

As Lingis paraphrases Deleuze's virtual ballistics, experience is dependent on the limitations of the system, but we are given the impression of potentialities: "This system of possibles cushions the impact of reality; without it the real abruptly turns up before my eyes to assault me like a blow struck from the invisible."[93] Of course, this does not shelter us from shock, or from the effects of participation that are alluded to, to a certain degree, of other different media platforms.[94]

Algorithm-based targeting and hitting depends on the screen as interface, which is how Virilio reads a particularly modern representation of movement: that of the car. The car seems to synthesize both ways of interpreting the game: screen/body supremacy in relation to spatial interpretation. Another analogy could be traced here—between the computer screen and the windscreen or windshield of a car. This analogy structures the ideological regimen of new spaces or new viewing experiences organizing themselves without effective self-consciousness. In the spatial relation of game space and viewer agency, Galloway sees a philosophical metaphor of thought and how space is cultivated. Space is fascist, targeted, or neoliberal, working through the flux of organicity.[95] The viewpoint of the video game through the screen is transformed into a spatial metaphor for politics or philosophy. But whatever interesting spatial relationships created through the space on screen, this is all nullified critically by Bartholl: the cement concretizes the ominous relationship to a space that is only unconsciously experienced.

Virilio's domain is optics and ballistics. In looking more closely at the windscreen as the monitor screen, we find an intriguing passage from Virilio's *Negative Horizon*: specifically, his explanation of the concept of dromoscopy through the coupling of the dashboard and windshield. At first, Virilio contrasts *dromoscopy*, a display of "inanimate objects as they were animated by a violent movement," to *stroboscopy*, where rapidly moving objects are displayed in slow motion.[96] Then he explains the changes and successive movement of the landscape when it is behind the glass of the windshield. The hurling objects on the "layer of the windscreen" are soon ejected beyond the rear window. Virilio writes that through speed, the "ground of the landscape," or, better, the depth of the landscape,

> rises up to the surface, inanimate objects are exhumed from the horizon and come each in turn to permeate the varnish of the windscreen, perspective becomes animated, the vanishing point becomes a point of attack sending forth its lines of projection onto the voyeur-voyager, the objective of the continuum becomes a focal point that casts its rays on the dazzled observer, fascinated by the progression of landscapes.[97]

In front of Virilio's windscreen, the observer is a voyeur and a voyager: someone who sees for pleasure without being seen, but also someone who actively seeks new spaces to explore. We can also consider the video game player in these terms, but what Bartholl does is marry the body's movement with astonishment in front of the negative space, sublime as opposed to fascinated, drawn in, attracted to an always-moving landscape. Virilio sees this format in many other places: for example, painting. He himself seems to make connections to a more static and contemplative art form: before the windshield, the "director of the movement [sets] himself up before a sort of easel composed of the screen of the windshield and the control panel of the components of the motor-projector."[98]

The screen, like a video game or Deleuze's desert island, is a capture of violence. We see this even in *Friday* when Robinson is confronted with the awestruck goat, which he immediately sees and treats as a target. But for Virilio, the violence is not a one-on-one encounter; it is the confrontation of the bunker and the sea creating a duality between two contradictory elements: the monumental and the elemental. This is a different perspective of the sea than that of Sloterdijk's delimiting function: through contrast, the sea makes the bunker bizarre:

> This architecture's moderness was countered by its abandoned, decrepit appearance. These objects had been left behind, and were colorless; their gray cement relief was silent witness to a warlike climate. Like in certain works of fiction—a spacecraft parked in the middle of an avenue announcing the war of the worlds, the confrontation with inhuman species—these solid masses in the hollows of urban spaces, next to the local schoolhouse or bar, shed new light on what "contemporary" has come to mean.[99]

In fact, the effect of Bartholl's piece is similar. Virilio is concerned here with the significance of this "other" architecture. We are switching scales from the individual to the architectural. The other, which makes us see the path in an isolated space, is included in the architectural material. Its specific function translates uncertainly into Virilio's contemporary situation. The very fabric of reality is torn: "The antiaircraft blockhouses pointed out another lifestyle, a rupture in the apprehension of the real."[100] The territory Bartholl's bunker-island occupies could be compared to the space between the virtual-real and the material-to-come—both of them somehow subverting the gamer's/viewer's relationship to real space. The bunker allure of *Dust* emerges from the virtual screen like a new reality, a potentiality. It is no accident that Bartholl's projected monument has connotations of the bunker and all of the psychological associations that come with it as a concrete manifestation of war. This is why even though Bartholl appears to celebrate the space of the video game, he also seems to make it into a funerary monument for the violent impulses that make the video game so appealing.

As the next section will demonstrate, the materialization of space we have seen with Bartholl is staged in Ridley Scott's *Prometheus* in a key scene featuring a three-dimensional holographic map, which is generated to show the chthonic bowels of a labyrinthine system of tunnels. But in the movie we see a reverse function to that in Bartholl's process—or a variation at least: the negative space of an invisible structure is virtually mapped out in positive terms. Whereas *Dust* is a solid manifestation of a hollow virtual space, *Prometheus* will offer a full virtual space representing an off-screen hollow but material space. Scott's will be the third manifestation of the island, creating a circuit between three map-images of islands constituting a visual apparatus. The first island, a document within the space of the Internet that comments on the selling of the desert island dream, connects with the second island, an unfinished art project that gives volume to a virtual game space where millions congregate through their data-selves, and finally, the island in *Prometheus* will be a visual model connected to Bartholl's artwork, but also the film itself will be a retelling of the desert island myth presented by Kerbel earlier: but here, the castaway will see the utopian island dream evaporate.

Prometheus

The idea of the other on the island is nowhere more pronounced than in Ridley Scott's *Prometheus* (2012)—except, of course, in its originating *Alien* (1979).

Prometheus is a cartographic adventure that situates the desert island trope in reverse: the planet, or island, on which the crew of an exploration spaceship lands receives many soon-to-be castaways (much is made of mission director Meredith Vicker's (Charlize Theron) private quarters/escape pod), but after a series of encounters, only one living human remains—a castaway by atrophy.

Prometheus has not been judged as Scott's masterpiece; critics found it confusing and rudderless. For my purposes, though, this rudderlessness is in fact a bonus element of the film. Walter Benjamin's theories on allegory advanced in his book on the *Trauerspiel* are certainly in line with *Prometheus*—as is this German tragic form in its own right, in which (almost) all the characters die at the end. Furthermore, the artificiality that rigidified the form of the seventeenth-century plays that Benjamin studied is present here: the narrative is stretched and compromised to fit into a series of external narratives and to bend to material pressures. The integrity and self-reflexive limitations of the artistic framework are secondary considerations in the overall object(ive) of the film. This is in keeping with postmodern aesthetics, especially those articulated by Lyotard (eclecticism),[101] Owens (accumulation),[102] and Jameson (the vagueness of signs)[103] in various texts at different times. If the film is an accumulation of images, a heaping of material with no regard for structure or meaning, then studying maps within the film is a delirious way out of the post-postmodernism. The maps offer a clear trajectory, a directionality not only as a cumulative Ariadne's thread within the plot, but also in themselves as directional objects reflecting upon the spatiality of the image in which they are framed.

In one sequence of *Prometheus*, characters enter a cavernous space for the purpose of geological mapping: an obscure and tortuous space of what appears to be an underground, crypt-like structure is later exposed as an alien weapons storage facility. The geologist mapping the network of tunnels gets lost, revealing the existential pitfalls of cartography as a cinematic device and the subject of the film: "*Prometheus*, as often in science fiction (e.g., *2001: A Space Odyssey*), maps a journey from scientific exploration to questions about our very nature that go beyond the reach of measuring tools."[104] This soon-to-be-lost geologist exploring the subterranean terrain releases his measuring tools, in this case, drones, which—shooting red beams onto the walls—take off down the corridors. They transmit data back to the ship, where the crew observes the generation of a three-dimensional map projected above a horizontal screen—in effect a screen-table—in the control hub. But even though the map seems to be helpful and its projection will clarify the darkness the crew has stepped into, it is a ruse on the part of the filmmaker: the official-looking diagram in fact does not give us the whole story. Through its deliberately clear, austere, and transparent network of lines, it actually camouflages a significant part of the information regarding the series of tunnels and networks the exploration crew has entered. The revelation of this element, obscured in plain sight, will create a sense of urgency and precipitate whatever is left of the crew to a more definitive extinction.[105]

The maps in the film are the propulsion engine of the plot. In fact, as Conley states in *Cartographic Cinema*, maps are tautological to film: "Even if a film does not display a map as such, by nature it bears an implicit relation with cartography."[106] The function of the map is to propel the narrative[107]—as even when the map is not clearly in view in a film, the screen behaves like a map: "A film can be understood in a broad sense to be a 'map' that plots and colonizes the imagination of the public it is said to 'invent' and, as a result, to seek to control. A film, like a topographic projection, can be understood as an image that locates and patterns the imagination of its spectators."[108] As we have already seen, Conley is not writing about typical geographical maps. In this way, *Prometheus* is a map in the *Alien* universe. A prequel, or somewhere in between, it becomes a nexus between the temporal lines of past films and the potential new ones. It awkwardly straddles traditions, but also tries to set up new elements without giving away too much: hence, the baroque sequel-machine, awkward and wooden, like a shabby seventeenth-century trauerspiel. Rüdiger Heinze, in an article titled "This Makes No Sense at All," discusses the many levels of narrative contexts that need to be taken into consideration when watching *Prometheus*. The result of a piling-up effect of "storyworld upon storyworld," constellations and fictional universes, is a film heaving with references awkwardly fitting together.[109] Under the weight of such cartographic manipulations (worlds, universes, constellations) the film becomes, according to Vivian Sobchack, "incoherent as narrative."[110] She borrows Benjamin's concept of dialectical images, seeing the confusing construction of the film's narrative as something more akin to a spatialization of film, in effect stalling the action onto a static dimension, a map, a "tension-filled constellation."[111] The overlaying effect is underscored by a sequence in which the explorers in the tunnels are overwhelmed by running phosphorescent aliens, a projection of surveillance footage documenting the demise of past inhabitants of the facility. Two paths, virtual and actual, crossing in a singular place. Two superimposed slices of time, as Deleuze formulated it in relation to Resnais's topological cinema.

Prometheus is a problematic map outwardly, materially, in the universe of the franchise as well. In this case we can perhaps also consider the film as allegory in the Benjaminian sense of piling on images. For Conley, the film's cartographic element does not simply orient us with regard to the movie within a franchise universe but is something much more existential: "A map in a movie begs and baits us to ponder the fact that *who we are* or whomever we believe ourselves to be depends, whether or not our locus is fixed or moving, on often unconscious perceptions about *where* we come from and may be going."[112] In this case, *Prometheus* becomes a tautological allegory of a map. Trying to find the source of life navigating toward origin, the character of Elizabeth Shaw, played by Noomi Rapace, goes further back to the source of death.

The importance of Conley's text is the shift from map to diagram. In this case, the para-filmic accessories of the film or its archival aspects—charts, maps, documents that are featured in the film—constitute a diagram of the film. They serve as traits that give rise to an actualized series of images; they provide directionality. Conley explains how "the sight of a map in a film often makes visible the history of the form producing the film, in other words, the archive held within and generating the tactics of the diagram."[113] In order to fully understand the relationship between map, diagram, and archive, one has to see the relationship between the three terms in *Foucault* by Deleuze. Conley establishes how these three elements within the film create something beyond the film itself: "As a 'diagram' or a model that maps perception and comportment through the image-field, the map is in flux where it shows how the archival aspect of the film might also be its diagram. The fluid and shifting space of the film and its cognition become *terrae incognitae* that the viewer explores in different directions and from various angles."[114] In *Foucault*, Deleuze exposes the logic of the diagram as a series of traits of a function that migrate from one system to the next, and how Foucault, as an archivist and cartographer, articulates cultural objects by tracing the functions of systems and mapping out the traits' trajectories. But all of these elements are stratagems to harness the force that comes from the outside: thinking, which takes place between milieus and which, outside the diagram, shapes them. At the end of *Foucault*, Deleuze draws us a stratigraphic map of this outside: representing the unrepresentable. The *terrae incognitae* that Conley sees in the transitions between diagrams, effectively working through the cartography of the screen, is this dimension, this outside, which can be captured by a fluid, shifting, malleable map-image.

Prometheus is, in fact, a film about three types of maps: archaeological, three-dimensional, and immersive.

Archaeological map

The first map is found in the cave at the beginning of the film: it shows an enigmatic pattern of circles above prehistoric representations of human figures. Later in the film, when the archaeologists summarize their discovery, they show how the map they found, and the enigmatic pattern, are replayed throughout ancient representation across cultures and across art history. These maps are central to the plot—in fact they are its main propulsion. As an antipathetic geologist, Fifield (Sean Harris), addressing Elizabeth Shaw and Charlie Holloway (Logan Marshall-Green), the archaeologists who instigated the mission through their discoveries, states: "So you're saying, we're here because of a map you two kids found in a cave? Is that right?" He means to be dismissive, but his statement is actually underscoring the importance of the map for the journey. The reliefs represent several cultures: Egyptian, Babylonian, Sumerian, Mayan, Hittite, Hawaiian, and a prehistoric cave painting from the Isle of Skye. Ranging in years from 35000 BCE to 620 CE, the papyrus, reliefs, stone carvings, cave paintings, and tablets all carry the same pictogram of figures turned upward to the sky. The six-dot pictogram makes up the planetary systems that the archaeologists interpret to be an invitation from the Engineers, extra-terrestrials ostensibly credited with the creation of human life on earth.

Three-dimensional map

The three-dimensional map is of central importance to the film. The holograph of the alien bunker tunnels displayed on the ship's control deck's digital table is almost a character in its own right: "The display of this information is rich with a saturated-color, color-coded, edge-opacity style, leaving outer surfaces rendered in a gossamer cyan, and internal features rendered in an edge-lit green wireframe. In the area above the VP [Volumetric Display] surface, other arbitrary rectangles of data can be summoned for particular tasks, including in-air volumetric keyboards."[115] The holographic map, which looks like a green island with a single mountain or volcano—an island-cliché, as it were—

is the result of the charting made by drones inspecting every surface of the chthonic territories. The three-dimensional green hologram that appears at the beginning of their exploration resembles bowel-like tunnels in their negative form.

Holographic maps—*holos* coming from whole or complete and *gramma* as something written—were developed as "photographic film or plate produced using a split laser beam that allows the viewer to see a three-dimensional image when viewed in coherent (later) or ordinary light."[116] This suggests that such a holographic map will give you a whole picture; in the case of *Prometheus*, however, the map is a decoy, since it essentially covers up more than it shows.[117] But the map in the film also resembles military grade maps developed since 2005: the monochrome green is an iteration of those maps, having the advantage of showing not only buildings in elevation but also what is underneath the terrain (gas mains, sewers, etc.), and also containing annotations.[118]

Hollowness appears as solid space in *Prometheus*. At 55 minutes 41 seconds, the viewer gets a better idea of what the drones have mapped [*Figure 17*]. The overall topography of the space appears as a mound, or a volcanic island, in effect a mountain on a desert planet, outlined in gray. The inside structure is at first glance a spiral, recalling Leonardo da Vinci's *Aerial Screw* (the connection suggested perhaps by the *Vitruvian Man* already making an appearance earlier in the film through "arbitrary rectangles of data"), or perhaps the floor plan of the Guggenheim Museum with its six rotunda levels. It will turn out to be closer to Botticelli's map of Dante's *Inferno* (1480–95) as the crew meets its demise. Such films as *Prometheus*

> display technologically advanced computer screens with which characters interact in order to access data and visual information. Normally holographic in nature, the screens project data with a translucent quality that works effectively in negative parallax. In a similar manner to stereoscopic debris, the non-solid appearance of these screens allows them to float effectively in the auditorium space. At times, their transparent nature makes it impossible to determine whether they are resting on a specific spatial plane. At other times, because their transparency allows objects behind them to be seen, a multitude of depth planes become apparent at the same time. The stereoscopic effect enhances this layering quality and brings into play overt relational depth.[119]

This map, a product of field research in the strange structures the crew explored, is featured as the focus of the command center of the ship. It is there as an objective fact of the "what" the crew is exploring away from the ship.

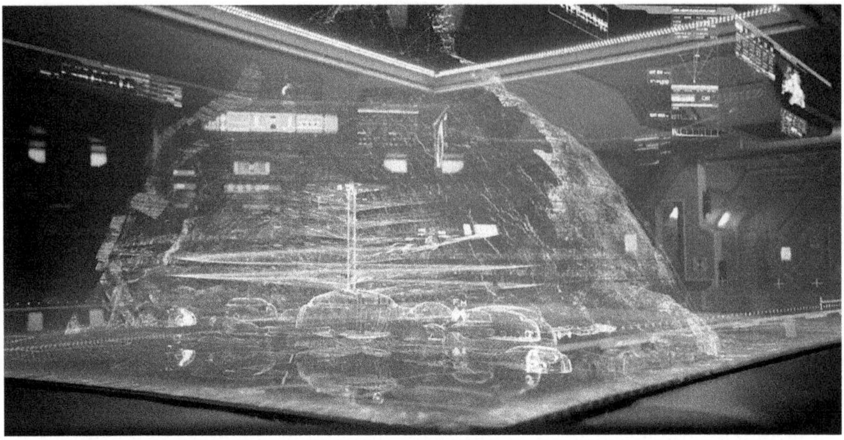

Figure 17 "PROMETHEUS."

We can read how this holographic, three-dimensional, dynamic, and incandescent map came to be through Deleuzean cartographic theory provided by Conley. When the ship's crew finds itself in the dark bowel-like corridors of the mysterious structure they have found on this new planet, the geologist Fifield releases two spherical mini-drones—or "pups" as he calls them—into the air. He communicates with the ship's captain, Janek (played by Idris Elba):

Fifield: "Prometheus, we're mapping."
Janek: "Copy that."
Janek: "Fifield, I got a read."
Fifield: "Yeah, Pups are saying: this way" [howls].

This exchange immediately brings to mind the map Conley examines in *Deleuze Cartographe*. The "map" Conley analyzes is, in effect, a photograph of wolf tracks in the snow serving as frontispiece to the "1914: One Or Several Wolves?" chapter of *A Thousand Plateaus*. Fifield actually howls, which signals his becoming animal—realized later in the narrative when his body undergoes a series of plicatures: inhabited by an infective alien organism, his skeleton folds from the inside and gives his body a crustacean gait. The exchange between Fifield and Janek also brings to the fore the duality between mapping and decalcomania that is the central element of Deleuze and Guattari's cartographic principle. If the map traces countries unknown, it must somehow copy in order for them to be recognizable. The problem of representation between the unknown and the familiar will be made apparent by this very map.

Just like the first kind of map, which traces the trajectory and therefore is the sole originator of the plot of the film, this table top-ological projection is also plot generating and revelatory of the narrative: At 1:36:12, Janek asks for the schematics to be put on the table. He asks that the dome be stripped away and that an area the viewer does not see be isolated. The isolated area appears as a floating blue and green crescent-shaped island. But after the area is enlarged and rotated, it is revealed to be a "goddam ship." What we thought to be a topographical map of the mountainous site, the scan of a pyramid, or an architectural plan of the internal structure is in fact an apparatus holding a spaceship. The map is the generator of the climax.

I will briefly mention three more points on the three-dimensional map before turning to the next. First, the crescent shape of the spaceship is reminiscent of More's Utopia (and by extension Kerbel's *Bird Island*), with definitive dystopian attributes: it is meant to travel to Earth to destroy mankind. Second, it seems that here we are back to Sloterdijk and his capsules: in a pointed fashion, the capsule contains the world and the universe. And third, the rendering of the topographical map of the inside of the earth in three dimensions is certainly reminiscent of Bartholl's *Dust* materialization.

Immersive map

In the Engineer's ship, the android David (played by Michael Fassbender) discovers an immersive, three-dimensional map plotting a route to earth (an intergalactic GPS with a Pepper's ghost–type of interface).[120] An obscure mechanism triggers the apparition of a luminescent map of the universe. It is a repeat trip: they are set to destroy what they created there. This is a reverse Robinson Crusoe trip: finding a populated island in the cosmos and rendering it deserted. The map here, despite its clearly enchanting appearance, showing the undiscovered beauty of the vast universe, has decisively dystopian overtones.

The aesthetic Pepper's ghost-like look of the map was used earlier in the film for specter-like apparitions of aliens running past the exploration crew in the tunnels of the facility. In this case, phosphorescent voxels were used to indicate a projected recording:

> The other alien VP tech is made up of small, blue-white voxels that float, move in space, obey some laws of physics, and provide a crude level of resolution. These appear in the caves of the alien complex where display tech is not present in the

> walls, and again as "security footage" in the bridge of the alien ship. Because the voxels obey some laws of physics, it's easier to think of them as glowing bits of pollen.[121]

We are back to the phosphorescent people, a diagrammed notion of artistic potential of the desert island. But in *Prometheus* this has been turned on its head: the phosphorescent beings have succumbed to their own destructive creation. They are, in effect, the pollen-like phosphorescent people that Lingis and Brito were seeing as populating deserted islands but with the dystopian consequences of a destructive bio-technology. In some ways, we are presented with a desert island at the end of the film. Shaw and the android have left in a capsule, and we can imagine that the phosphorescent people will perpetually enact their demise on the desert island planet.

Conclusion

The image of the island is a *dispositif*. Giorgio Agamben explains in "What Is an Apparatus?" the function of a *dispositif*: the apparatus, originating in Foucault, is defined as, among other things, a network of relations.[122] Agamben, after having established the network element of the term "apparatus", proposes the following way of looking the concept: "The term, 'apparatus', designates that in which, and through which, one realizes a pure activity of governance devoid of any foundation in being. This is the reason why apparatuses must always imply a process of subjectification, that is to say, they must produce their subject."[123] For Agamben, there are two large groups of beings: the ones that are living and apparatuses ("in which living beings are incessantly captured")[124]—ontology and economy. In this way, an apparatus is anything that has "the capacity to capture, orient, determine, intercept, model, control, or secure the gestures, behaviors, opinions, or discourses of living beings."[125] This sounds like Parikka's ethology,[126] or Krauss's etiology,[127] or perhaps even Galloway's ethics (politics)[128] of the digital. Admittedly, the term *dispositif* requires some kind of adaptation in order to fit the aesthetic image of the island we have been investigating. The image-island here is not a mere representation: it is an idea, a model, a concept, an image, a plan, a website, a chart, a game space, a sculpture, a metaphor, an analogy, a fiction, a story, a visualization, a capsule. All of these manifestations are different facets of capture of the living: the castaway, the upcoming people, the websurfer, the island-dreamer/investor, the gamer, the art-aficionado, the moviegoer, the space crew, the phosphorescent people. The island-image is self-reflexively shuttling between the outside and the inside of the work of art. But in keeping with the posthuman or nonhuman aspects of digital intermediality that we have already explored in the past, this is something Agamben also sees as an issue of interface with new modes of desubjectification: "But what we are now witnessing is that processes of subjectification and processes of desubjectification seem to become reciprocally indifferent, and so they do not give rise to the recomposition of a new subject, except in larval or, as it were, spectral form. In the nontruth of the subject, its own truth is no longer at stake."[129] One of the ways that Deleuze will endorse this idea of desubjectification is through his concept of dividuation; Agamben himself warns of the danger of an apparatus of capture and control through a desubjectified individual.

All three objects—*Bird Island, Dust,* and *Prometheus*—are frameless. The island website is unattached, floating, it seems, in the vast Internet. Bartholl's work is not completed as I am writing this. And Scott's film is ill-fitting within the *Alien* universe. All three fleetingly capture a digital non-place: they are each an island-apparatus capturing digital ruins. Maybe the island offers the possibility of a vacuole of silence to break away from the regime of control, which, as Deleuze observes, is never done with anything.[130] The island becomes a space of solitude, interrupter of the control that can lead to desubjectivation through the island-image.[131] A *dispositif* as a cure for *dispositifs*. A switch of capture and release. And perhaps the island as interrupter breaks the circuit that constitutes an archipelago of the impossible, impractical, and destructive.

Notes

1 Deleuze and Guattari, *What Is Philosophy?*, 36.
2 Gilles Deleuze, *Logic of Sense*, ed. Constantin V. Boundas, trans. Mark Lester (New York: Columbia University Press, 1990), 309.
3 Deleuze, *Francis Bacon*, 159.
4 Deleuze, *Essays Critical and Clinical*, 119. See Tom Conley, "The Strategist and the Stratigrapher," in *Afterimages of Gilles Deleuze's Film Philosophy*, ed. D. N. Rodowick (Minneapolis: University of Minnesota Press, 2010), 210n32.
5 Gilles Deleuze, "Causes and Reasons of Desert Islands," in *Desert Islands and Other Texts: 1953–1974*, ed. David Lapoujade, trans. Michael Taormina (Los Angeles: Semiotext(e), 2002), 11: "Deserted, the island may be a desert, but not necessarily."
6 Conley mentions the geographic quality of Deleuze's philosophy.

7 Tom Conley, "The Desert Island," in *Deleuze and Space*, ed. Ian Buchanan and Gregg Lambert (Edinburgh: Edinburgh University Press, 2005), 217.
8 Ibid., 208, 209.
9 Ibid., 86.
10 Deleuze and Tournier in Deleuze, *Logic of Sense.*
11 Gilles Deleuze, "Causes et raisons des îles désertes," in *Île Déserte et autres textes: Textes et entretiens 1953–1974*, ed. David Lapoujade (Paris: Minuit, 2002), 12.
12 Ibid., 13.
13 Ibid., 10.
14 Ibid., 13.
15 Ibid., 11.
16 Ibid., 12.
17 Elizabeth Berkebile McManus, "Protecting the Island: Narrative Continuance in *Lost*," *Journal of the Fantastic in the Arts* 22, no. 1 (81) (2011): 15.
18 Alphonso Lingis, "Deleuze on a Deserted Island," in *Philosophy and Non-philosophy since Merleau-Ponty*, ed. Hugh Silverman (Evanston: Northwestern University Press, 1997), 154. "If the others as a category constitutive of the field of objective-objectives, what Deleuze calls the other as an *a priori*, which precedes and makes possible the encounter with any empirical other, is the possible and the future affixed to every present, then the experience of a world inhabited by other is the experience of a time in which each present jostles on the heels of another time." Ibid., 160.
19 Deleuze, *Essays Critical and Clinical*, 115: "Haze is the first state of nascent perception … To see through haze is to have blurred vision—the rough outlines of a hallucinatory perception a cosmic grey."
20 Gregory Flaxman, *Gilles Deleuze and the Fabulation of Philosophy* (Minneapolis: University of Minnesota Press, 2012), 85.
21 Ibid.
22 The paradoxical temporality of the time on the island could perhaps be explored along the lines of the virtual in cinema as worked out by Deleuze: "Thought remains forever outside the formulation where it takes on form. The force of the cinema invented by Deleuze would be to embody this double gesture of exclusion and attraction of a thought that is only to come, but that will come to be, by leaps and discontinuity, always ready to break with its own trajectory." Ropars-Wuilleumier, "Image or Time?", 28.
23 Ibid., 86.
24 Ibid.
25 Vanessa Brito, "L'île déserte et le peuple qui manque," in *Gilles Deleuze et les images,* ed. François Dosse and Jean-Michel Frodon (Paris: Cahiers du Cinéma Institut National de l'Audiovisuel), 72, my translation: "Il ne peut être qu'un peuple

moléculaire, un peuple phosphorescent fait de grains dansant et de poussière lumineuse qui était déjà là à l'origine de l'œuvre d'art: le peuple atomique de l'île déserte dont la danse se confond avec les combats mythologiques entre les éléments de la nature."

26 Ibid.

27 Flaxman, *Gilles Deleuze and the Fabulation*, 85.

28 Deleuze, *Essays Critical and Clinical*, 115.

29 Conley, "Desert Island," 212.

30 Daniel Smith and John Protevi, "Gilles Deleuze," in *The Stanford Encyclopedia of Philosophy*, ed. Edward N. Zalta (Winter 2015), available online: https://plato.stanfod.edu/archives/win2015/entries/deleuze/

31 Peter Hallward, *Out of This World: Deleuze and the Philosophy of Creation* (London: Verso, 2006), 23.

32 Conley, "Desert Island," 208.

33 Ibid.

34 Kim Dhillon, "See It Now or Miss It Forever: Materiality, Visuality, and the Written Word in Janice Kerbel's Recent Artwork," *RACAR* 36, no. 1 (2011): 24.

35 Isolating: As Jon Ippolito explains in "Ten Myths of Internet Art," *Leonardo* 35, no. 5 (2002): 485–87, 489–98.

36 Kerbel's Gardens are meant to be commentaries on climate change and urban habitat. She writes: "Even under threat, nature continues to be seen as a metaphor," and as a metaphor, nature keeps giving us a potential model for thought. Kerbel explains how the contradictory attitude toward climate change is based on the push of utopian desires and the pull of dystopian habits: "The natural environment is threatened by the built one." We are not able to imagine the environment, to figure out how to solve the problem of climate change. Even scientific data is up for debate. The question posed through Kerbel's aesthetic strategy centers on the figuring of the environment, giving it a shape and laying out a ground for action.

37 Krauss, "Stieglitz/'Equivalents,'" 133.

38 Maurice Blanchot, *The Infinite Conversation*, trans. and forw. Susan Hanson (Minneapolis: University of Minnesota Press, 1993), 309.

39 Ibid.

40 Ibid., 289.

41 Sloterdijk, quoting Georg Simmel, in Peter Sloterdijk, *Spheres Volume 3: Foams: Plural Spherology*, trans. Wieland Hoban (South Pasadena: Semiotext(e), 2016), 289.

42 Sloterdijk, *Spheres Volume 3: Foams*, 295.

43 Ibid.

44 Ibid., 290.

45 Ibid.

46 Ibid.
47 Ibid., 296.
48 "It is imperative for the design of the absolute island that the loose atmotopic exceptional situations of natural islands be recreated in the strict exceptional situation of the artificial closed atmotope." Ibid., 298.
49 The idea is entering the realm of the actual these days: see "For Libertarian Utopia, Float Away on 'Startup' Nation," *Information Liberation*, 2 June 2014, available online: http://www.bloomberg.com/news/articles/192014-05-30/for-libertarian-utopia-float-away-on-startup-nation
50 Anna Lovatt, "Janice Kerbel: Diagramming Desire," *Afterall: A Journal of Art, Context and Enquiry* 37 (Autumn/Winter 2014): 90–101.
51 Elizabeth Grosz, *Architecture from the Outside: Essays on Virtual and Real Space* (Cambridge: MIT Press, 2001), 135.
52 Ibid., 136.
53 Ibid., 138.
54 Ibid., 143.
55 Thomas More, "Utopia," in *Three Early Modern Utopias: Utopia, New Atlantis, The Isle of Pines*, ed. Susan Bruce (Oxford: Oxford University Press, 1999), 49.
56 Ibid.
57 Jules Verne, *L'île à hélice* (Paris: Union Générale d'Éditions, 1978), 30.
58 Jules Verne, *Propeller Island* (London: Granada Publishing, 1977), 24.
59 Hubert Damisch, "Genealogy of the Grid," in *The Archive of development*, ed. Henk Slager and Annette W. Balkema (Amsterdam: Rodopi, 1998), 51.
60 Ibid., 54.
61 Ibid.
62 Ibid.
63 Verne, *hélice*: "De même pour les squares, ménagés à l'intersection des principales artères de la ville, tapissés de pelouses d'une fraicheur tout anglaise, dont les massifs, où se mélangent les essences des zones tempérée et torride, n'ont pas aspiré des entrailles du sol assez de puissance végétative." My translation.
64 "Aux amateurs de peinture, le musée, riche de tableaux anciens et modernes, offre de nombreux chefs-d'œuvre, acquis à prix d'or, des toiles des écoles italienne, … hollandaise, allemande, française, que pourraient envier les collections de Paris, de Londres, de Munich, de Rome et de Florence, des Raphaëls, des Vincis, des Giogriones, des Corèges." Verne, *L'Île à hélice*, 75.
65 Curiously, Verne published the book and the Italian futurists published their manifesto in 1909.
66 "Grâce à ce climat sans pluies ni brouillards, groupes, statues, bustes peuvent impunément résister aux outrages du temps." Verne, *L'Île à hélice*, 75.
67 Deleuze and Guattari, *What Is Philosophy?*, 163.

68 Martin Heidegger, "The Thing," in *Poetry, Language, Thought*, trans. Albert Hofstadter (New York: Perennial Classic, 2001), 164.

69 Ibid., 166.

70 Ibid., 167.

71 Ibid.

72 Ibid., 169.

73 Bartholl artist's website: http://www.datenform.de/dust-rhizome-eng.html

74 Richard Shone, "Rachel Whiteread's 'House,'" *The Burlington Magazine* 135, no. 1089 (1993): 838.

75 Ibid.

76 Bartholl's website.

77 Ibid.

78 Ibid.

79 Ibid.

80 Marc Augé, *Non-Lieux: Introduction à une anthropologie de la surmodernité* (Paris: Éditions du Seuil, 1992), 101–2.

81 Ibid., 102.

82 Michel Foucault, "Of Other Spaces," trans. Jay Miskowiec, *Diacritics* 16, no. 1 (1986): 23.

83 Ibid.

84 Bartholl's website.

85 Deleuze, "What Children Say," 66.

86 Ibid.

87 Galloway, "*StarCraft*," 88–89.

88 Ibid., 89.

89 Ibid.

90 Deleuze, *Logic of Sense*, 305.

91 Deleuze, "What Children Say," 66.

92 Ibid., 67.

93 Lingis, "Deleuze on a Deserted Island," 156.

94 "It is the others too that constitute one as a mobile light-source that circulates among objects. Without them there is not first a theater of objects into which one watches oneself enter; there is only a moving zone of light in which patterns shine with their own phosphorescence.... this phosphorescence in the sensory elements is extracted out of them and shut up in itself as a separate entity, one's subjectivity, by the dense-packed core of other, prone to riot." Ibid., 157.

95 Galloway, "*StarCraft*," 91: "If *Counter-Strike* is fascism, *World of Warcraft* is neoliberalism: the one pegs everything on force as an uncontaminated aesthetic act, optical or otherwise; the other thrusts the constituents of the system into a flux of organic struggle. If, as Benjamin wrote, fascism is an aestheticization

of politics (and therefore not, following his suggestion, a politicization of the aesthetic), is not neoliberalism that new global system that distributes control in such fine granularity so that all must struggle against all? It is not unreasonable to suggest that we have witnessed a general transformation from the one to the other in all spheres of social and material life: from the modern model of power in the 'fascistic' visual episteme to the contemporary model of control in "neoliberal" information systems."

96 Paul Virilio, *Negative Horizon: An Essay in Dromoscopy*, trans. Michael Degener (London: Continuum, 2007).

97 Ibid.

98 Ibid., 105.

99 Paul Virilio, *Bunker Archeology* (New York: Princeton Architectural Press, 1994), 12.

100 Ibid., 13.

101 Jean-Francois Lyotard, "What Is Postmodernism?" in *Art in Theory 1900–1990: An Anthology of Changing Ideas*, ed. Charles Harrison and Paul Wood (Oxford: Blackwell, 1996).

102 Craig Owens, "The Allegorical Impulse: Toward a Theory of Postmodernism," *October* 12 (1980): 67–86.

103 Fredric Jameson, "Utopianism after the End of Utopia," in *Postmodernism, or, The Cultural Logic of Late Capitalism* (Durham: Duke University Press, 1991), 154–80.

104 Danny Pegg, "Prometheus," *Journal of Religion & Film* 17, no. 2 (2013): 6–7, available online: http://digitalcommons.unomaha.edu/jrf/vol17/iss2/17

105 Of course, this led to the creation of the Internet meme: "Has 3D mapping and videolink to the ship—still gets lost."

106 Conley, *Cartographic Cinema*, 1.

107 Ibid.

108 Ibid.

109 Rüdiger Heinze, "'This Makes No Sense at All': Heterarchy in Fictional Universes," *StoryWorlds: A Journal of Narrative Studies* 7, no. 2 (2015): 76.

110 Vivian Sobchack, "Between a Rock and a Hard Place: How Ridley Scott's *Prometheus* Deals with Impossible Expectations and Mythological Baggage," *Film Comment* 48, no. 4 (2012): 34.

111 Ibid.

112 Conley, *Cartographic Cinema*, 3.

113 Ibid., 15.

114 Ibid., 21.

115 Christopher Noessel, "Bridge VP Mapping," *Sci Fi Interfaces* (blog), 28 November 2012, available online: https://scifiinterfaces.wordpress.com/2012/11/28/bridge-vp-mapping/

116 Eduard Hölzel, "Holographic Map," in *The History of Cartography, vol. 6, Cartography in the Twentieth Century. Part 1*, ed. Mark Monmonier (Chicago: University of Chicago Press, 2015), 619.

117 Ibid.

118 Zebra Imaging has been developing holographic maps that Hölzel describes as "generated three-dimensional holographic maps that could be illuminated under battle field conditions or during an environmental disaster using a portable light or even sunlight. Rendered in green monochrome or tru color, these maps could show underground features like gas mains or sewers, using dry-erase markers or grease pencils. Proposed uses included rescue and evasion, evacuation and recovery, military defense, and homeland security." Hölzel, "Holographic Map," 619.

119 Miriam Ross, "Stereoscopic visuality: Where Is the Screen, Where Is the Film?" *Convergence: The International Journal of Research into New Media Technologies* 19, no. 4 (2013): 410.

120 Christopher Noessel, "Alien VPs," *Sci Fi Interfaces* (blog), 18 December 2012, available online: https://scifiinterfaces.wordpress.com/2012/12/18/alien-vps/: "The two alien VPs are quite different from the human VPs in appearance and behavior. The first thing to note is that they adhere to the Pepper's Ghost style more readily, with glowing blue-tinted whites and transparency. Beyond that they differ in precision and implied technology."

121 Ibid.

122 Giorgio Agamben, *What Is an Apparatus? and Other Essays*, trans. David Kishik and Stefan Pedatella (Stanford: Stanford University Press, 2009), 3.

123 Ibid., 11.

124 Ibid., 13.

125 Ibid., 14.

126 Parikka, "Ethologies of Software Art."

127 Krauss, "Grids," 64.

128 Galloway, *The Interface Effect*, 24.

129 Agamben, *What Is an Apparatus?*, 21.

130 Deleuze, *Negotiations*, trans. Martin Joughin (New York: Columbia University Press, 1995), 179.

131 Ibid., 175: "The key thing may be to create vacuoles of noncommunication, circuit breakers, so we can elude control."

6

Surveilling Aesthetics: Waheed's Overhead Images and Farocki's Operative Image

At the beginning of map-image, we looked at Vermeer through Deleuze, searching the painted surface for a matrix of information. The map signified, from the start, a way of seeing the world beyond the frame. But it also registered as a flat object, coextensive with the canvas that depicts the map, a horizontal object hung on a wall. Bring the map down to study it on a desk and you bring the painting down with you. You have a catalog downward view, a "*katalogia*," a view from above.[1] It would be indispensable to look at another instance of this view from above and see how the shift occurs from an information image to a posthuman image, an image that no longer conveys information on data but that is meant to operate beyond human perception.

In this last chapter, I would like to look at two artists, Hajra Waheed and Harun Farocki. Both are interested in visual expression that comes from a technologically enabled and algorithmically elaborated image. Yet each artist adapts a different set of strategies to convey this new posthuman aesthetics. Even if they are both interested in the aesthetics of surveillance, Waheed is more hands-on and personal, whereas Farocki is detached and seemingly non-interventionist. Waheed's work links to the flatbed aesthetic. Collage and plans surface, and maps are interconnected and provide information in the tradition of Robert Rauschenberg's works as described by Steinberg. Yet, her subject matter is that of surveillance, control, and machine point of view. The kind of expanse Waheed puts on display recalls the issue of distance calculated by theorists of Vermeer. But here, the beyond does not register human physicality and perception. Waheed opens a way of looking at the digital era with an aesthetic that still includes a humanistic approach to the art object. This is why on some level she creates the bridge that allows us to demonstrate how the map-image is part of a digital apparatus even in an analog medium. But once I have shown the posthuman thread linking our aesthetic objects together, I will turn to Farocki's surveillance images, which seem to illustrate the path on which the map-image had already begun to proceed. Operating on the level of posthuman information, the image in Farocki—which unlike a map that does not register on the human eye—provides a seemingly unmanipulated machinic sequence. The films are a series of edited shorts of technologically influenced images, relying heavily on montage to create a sense of order in the sequence. And like Waheed, Farocki uses a type of collage technique—but instead of cut-out paper fragments, two screens or projections play off of each other—presenting disparate images side by side, recalling a stereoscopic vision apparatus that has been usurped for a posthuman functionality.

Returning to the beginning and looking at Steinberg's shifting point of view between the canvas and the screen, we can adjust the line of sight even further and look down: down on the desk or floor or map. Or, in the case of surveillance, on the ground. Waheed's work—information, seen from above—provides an example of these types of images. Her use of collage, I believe, best summarizes Steinberg's concept. The directionality of the viewer vis-à-vis the canvas, as mitigated by Steinberg's concept of the flatbed picture plane and applied to Waheed's work, emphasizes our contemporary visual world's concern with information. Her collages resemble Rauschenberg's optimistic series on Kennedy-era space travel (Rauschenberg, of course, to whom Steinberg dedicated his concept, was the originator of the flatbed picture plane). But with Waheed, we are looking at the darker consequences of contemporary global issues related to surveillance, global security, and satellite spying.

Waheed explores notions of personal, national, and cultural identities in a body of work that is deeply political. Working in the vein of the Atlas Group, she weaves narratives both fictional and actual in order to bear witness to situations of which we cannot always have a clear view. This tentativeness of knowledge is in line with using the archive to subvert it. The *KH-21* (2014–15) series [*Figure 18*], for example, explores the declassified spy satellite program that Waheed displays through sculpture, drawing, and collage. By using archival material in an aesthetic context, she creates an oscillation between truth and fiction: the archive stands in for authenticity in the same way that the idealized photograph stands in for unwavering documentation. A relationship exists between documented images and apparatuses of power "in the form of organizational rules and archival protocols, as other spheres of documentation."[2] Archival "complicity with existing power structure" is an inherent element in Waheed's work, since

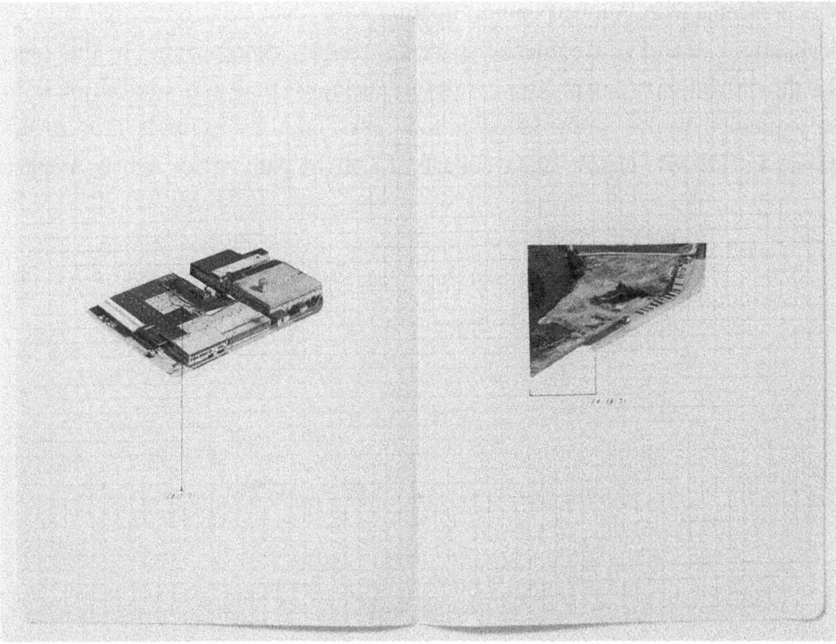

Figure 18 Hajra Waheed, *KH-21, NOTES*, 22/32 (2014). Cut Photograph and ink on paper. Courtesy of the artist.

the only reason *KH-21* was possible was because the US government declassified the data of its spy satellite program—the Hexagon program—which was part of the National Reconnaissance Office in the United States and their satellite launches in the 1970s and 1980s.

Waheed's work is more than just "a postmodern exercise marked by a preoccupation with the document and the documentary in relation to problems of truth and fiction."[3] *KH-21* deals with a new kind of informational image. The documentary aspect of Waheed's art is the scaffolding upholding her desire to document the "ever-growing vertical occupation," through a rumination on drones and spy satellites and the subsequent images that come out of this particular type of technology. What is interesting for the artist is to work with "undisclosed documents," which she recombines in "often sparse and centered [compositions] with images and text arranged as if in code."[4] Her work is firmly rooted in the spirit of documentation: she is "driven by a tendency to catalog, categorize and re-organize selected fragments."[5] The cataloging is key here, signifying not only the collection, organization, and classification of data, but also the downward movement etymologically hinted at by the term *katalogia*, which chimes with Waheed's preoccupation with "data collection, high altitude occupation … and … decades of space-debris now descending."[6] In this case, the downward-cast perspective results in an image that may not be made to be witnessed by the unmediated human eye. Take for example *Fear Brings About 4/5* (2010) [*Figure 19*], a collage of a flight path, space debris, a space

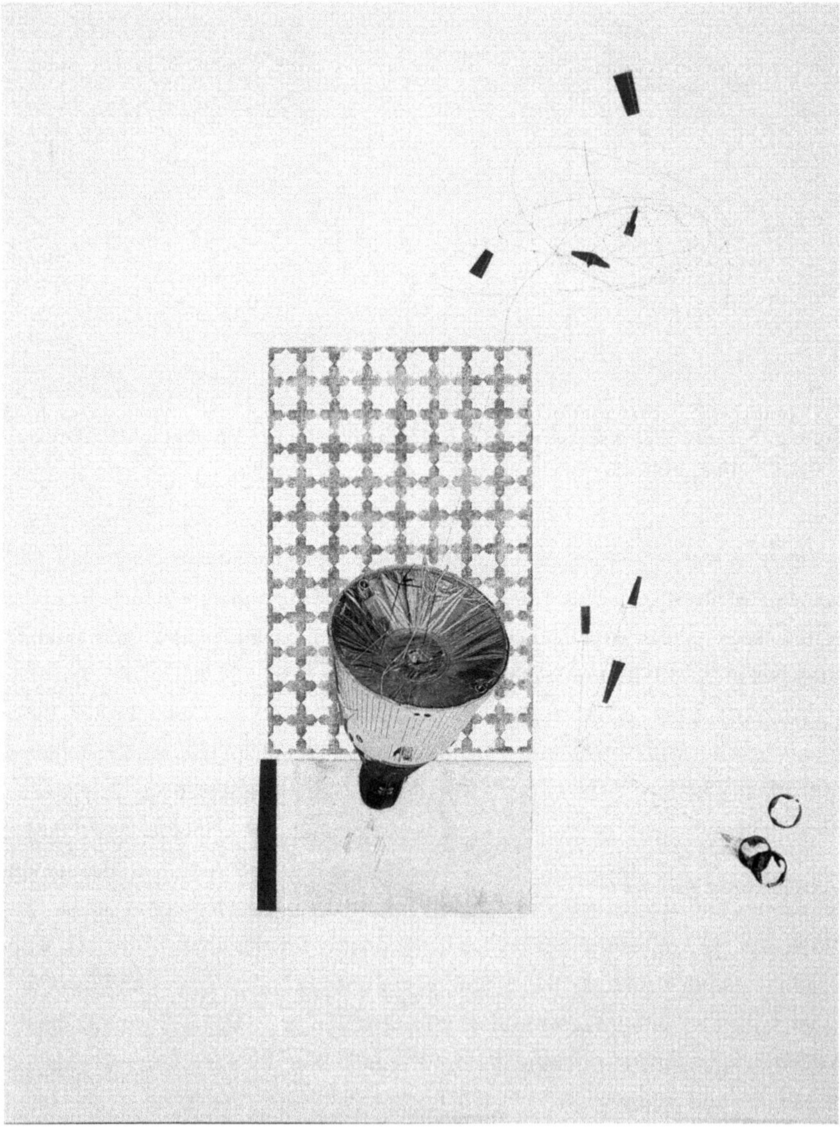

Figure 19 Hajra Waheed, *FEAR BRINGS ABOUT*, 4/5 (2010). Collage, acrylic, ink, and pencil on graph paper. Courtesy of the artist.

capsule, and a mosque floor plan that articulates notions of falling: overhead aesthetics and architectural elements become real poetics of posthuman technological culture. Lisa Parks, in her work on surveillance satellites, draws attention to the elusiveness of information with images of this type. She calls this image the "overhead image," defining it as "image-data that has been acquired by instruments onboard aircraft or satellites, downlinked to earth stations, rendered by computer software, and, in some cases, composed for the purpose of representation, viewing, and analyzing particular sites or activities on earth."[7] This visualization of data mediates information in such a way as to make it recognizable to individuals, who will then be tasked with their interpretation. We could say that Waheed mediated the source images even further, making them resonant as art bearing witness to the notion of mediated information and visualization of heretofore nonrepresentational images. What Parks stresses in her assessment of these "overhead images" is their invisibility—they are not meant to be disseminated, but are produced by a network that thrives on discreetness.[8] From an aesthetic perspective, these "overhead images" allow us to think about communication that is not solely representational,[9] "situated beyond human perception."[10]

Waheed's works are imbued with the personal touch of material manipulation, drawing, collage, and an individual sense of place and displacement. The fact that her exploration of surveillance cartographies has taken up residence at her psychic address, to use Steinberg's term, demonstrates a deep incursion into surveillance geography of a control society. More than anything, she exemplifies how the view from above is an extension of the flatbed point of view we have been using as a pivot point between illusion and information. But here, the scale has expanded beyond the human posture vis-à-vis a painting on a wall, beyond the hunched over posture of reading at a table: we are kilometers above the earth. Waheed's work is used as a emblem of sorts for a surveillance society because it illustrates art's strategies of dealing with the posthuman stature of images.

The posthuman thread running throughout this book registers at varying frequencies. In the beginning, it is noted in the qualities of code; then, in the leveling of information into a non-meaning "noise"; the virtual in its seemingly agentless potentiality also sidles up to the posthuman; and finally, we imagine the desert island turned dystopia erasing the figure of the human from the sand, as Foucault announced at the end of *The Order of Things*, "like a face drawn in the sand at the edge of the sea."[11] As we saw in the second chapter, Francine Savard's paintings negotiate between information aesthetics and Emmanuel

Lagrange Paquet's maps make for a different kind of antiquating rune out of the patterns of video game controller buttons. Coding for these two artists is embodied differently, and materiality appears as different iterations of a single informational stream. Issues of what is visible and turned into aesthetics of the code come to the fore. What begins rather low-key—a change of the guard, as it were, in terms of how art manages cartography—is reinforced in the third chapter through the art of Cory Arcangel. English is scrambled through English, picture is scrambled through lines, and even a blank screen becomes glitchy in order to show the human aspect of code. But what also becomes apparent are the distances between individuals in the global village and the advent of data-selves as a form of subjectification, the aestheticization of street codes through Hollywood's whitewashing, through appropriation of culture that renders any narrative pastiche. As for the white screen of the *Structural Film*, its fragility and deterioration seem to echo the map in Borges's story lying shredded, information lost. John F. Simon Jr.'s work seems to be quite colorful and pleasant and so comes close to the candy-colored lines that could mask the chasm between information and its visual representation. But Simon's works are animated by algorithms: cartographies of abstractions, like thought patterns as swarms, or at least the swarm-image that infects our thoughts; maps of ecosystems that can go on in perpetuity without any humans; or even fully animated abstracted cities, repeating their routines infinitely but with slight variations that provide a kind of life to each map.

Furthermore, the grid element in Simon's work conveys a necessary artificiality. We already see aspects of the grid in Deleuze's move from painting to screen and in Simondon's environmental virtual grids. The grid also made its appearance in the visual organization of aforementioned artists' works: Savard's texts and Lagrange Paquet's cosmic meshes, Arcangel's stripes, and even Vermeer's perspective matrix. On a more matter-of-fact level, there is something foreboding about the grid. It is the matrix upon which space is created. The grid pertains to the posthuman; however, as Krauss explains, it is a question of a spiritual or scientific beyond. But also, from the very beginning, the blind and dumb aesthetics of silence and repetition is apparent:

> The barrier it has lowered between the arts of vision and those of language has been almost totally successful in walling the visual arts into a realm of exclusive visuality and defending them against the intrusion of speech. The arts, of course, have paid dearly for this success, because the fortress they constructed on the

> foundation of the grid has increasingly become a ghetto. Fewer and fewer voices from the general critical establishment have been raised in support, appreciation, or analysis of the contemporary plastic arts.[12]

Simon explores this silent relationship between thought and information: the aesthetic of the posthuman. His *Every Icon* will only appear long after the sun explodes. And here is the fear of the algorithmic lack of agency that seems to present on the virtual, inaccessible to our senses and metaphorically expressed through Marks's algorithm/image dichotomy. The vistas presented by Simon are posthuman to varying degrees: animal swarming patterns that are meant to reflect thought—perhaps not necessarily human thought. These vistas are autopoetic ecosystems that function beyond human perception but that are nevertheless at its basis, even if imperceptible to consciousness. And they are cities, gridded since the invention of the brick; but as of late the gridiron seems beyond the individual human scale, and the fact that the patterns of the city are dependent on an algorithm provides a colorful visual allegory of a control society.

In the previous chapter, we appeared to have moved beyond the "exclusive visuality" of the grid, the abstract, and the virtual by focusing on the ostensibly pastoral image of idealized island. But, it is through the island that Kerbel puts into perspective the abstract vastness of the Internet and the egoistic desire to possess that which disappears once it is within grasp: the desert island that can never be deserted even when it is fictitious. The islands in *Utopia* or *Propeller Island* are like Kerbel's potential resort island, never reachable and yet always expressing a perverse desire to possess through order and administration—and even the grid structure makes an incursion, as is the case with Verne. Reading it through Sloterdijk's isolation and atmospheric manipulation of the artificial island removes us from Deleuze's intellectually idyllic island full of creative potentiality. Bartholl's island will be made of concrete eventually. It started off as a virtual patch in the desert. Here we are confronted with violence, through the screen as interface and the negotiation between the virtual and the actual. This set of negotiations for a new spatial aesthetics could be read according to Buci-Glucksmann's screenic virtual, where traditional human parameters are challenged by a new technological reality. And Scott's islands just seem to proliferate: the capsule, the space suit, the space craft. The desert island image in three-dimensions on a table—the island that changes into a capsule, bound for Earth in order to erase the mistake of humanity, erasing the face from the sand of the beach of the Deleuzean island.

What comes next is explained by N. Katherine Hayles in her *How We Became Posthuman*, where she offers a four-point definition of posthuman within a chapter that traces the evolution of the term parallel to that of cybernetics and its relationship to humanities. "First," writes Hayles, "the posthuman view privileges informational pattern over material instantiation," so that information is favored as the site of life over its biological embodiment.[13] Second, consciousness is but a byproduct, "an evolutionary upstart" that believes itself to occupy a position of privilege in relation to other models of intelligence.[14] Third, the body is a prosthesis, and therefore any technological augmentation to the body is simply part of its prosthetic program. Finally, "the posthuman view configures human being so that it can be seamlessly articulated with intelligent machines."[15] In relation to the images of mapping we have considered, what concerns me is how these facets of the posthuman are articulated spatially. Beyond active consciousness and within a network of technological extensions to the body, how do we map out space?

Perhaps Bruno Latour's view of the human without a stable form but nevertheless not formless, the human as mediator, would ground historically the inhuman "robot animated with neurons, impulses, selfish genes, elementary needs and economic calculation" as too reductive.[16] The human stands as part of a network: "*Ecce homo*: delegated, mediated, distributed, mandated, uttered."[17] This mediating, distributing, informal but not formless *anthropos* is at the crux of the mapping of networks, since, as "the weaver of morphisms" between nature and discourse, the posthuman eschews representation.

But how do we represent a relationship with the posthuman and the map? What is the posthuman map-image? Galloway, in his *Interface Effect*, attempts to discourage us from seeking to map information, such as the spatial charting of Internet connections. He decries the cauliflower maps that spring up all over cyberspace, carefully explaining the chasm between information and the look we want to give it, an incommensurability stemming from the fact that information has no aesthetic, inherently visual surface with which we could interface. Of course, then, the idea of spatializing information in order to quantify it is equally problematic. He writes: "Information interfaces, particularly the many attempts to 'map' information, often come up short on this score, for they typically offered little orientation within the social totality."[18] Here we can think of the work of Ryoji Ikeda, which captures the sense of awe that large amounts of data conjure in our imagination but then fails to demonstrate data's function and effect in society. Or the criticism often levied against Richard Ighby and

Marilou Lemmens, artists who work with the aesthetic element of information visualization without a hint of content, like in their work *The Prophets* (2013–15), which consists of a multitude of delicate and colorful infographic-based diagrams displayed on a series of tables. Galloway seems to have them in mind, even though, of course, he is speaking in more general terms, when he states that these attempts at mapping "often exacerbate the problem by veiling it beyond candy-colored lines and nodes."[19] For Galloway, this type of mapping is not obvious since it is the quixotic mission of capturing a totality of information, its abstract, disembodied location, and its relational functions. This leads him to state, "The tools and techniques required to create cognitive maps of the information society are scarcely evident even today."[20] But, Galloway suggests, an allegorical approach is permissible. The image of a chart detailing the US military strategy in Afghanistan in 2009 is actually a snapshot itself of how we imagine the information effectively functioning. For Galloway, the military map is a way of gaining access to an image of, and not the, actual mechanism of the operation: So the military image is not a map of a system but an "*allegory for* a map of a system."[21] Galloway says that "the difference is slight but crucial. … The point is not so much to call for a return to cognitive mapping but to call for *a poetics as such* for this mysterious new machinic space."[22] But even a poetics of cartographies of the digital yields a posthuman aesthetics, as Deleuze seemed to announce with his "Postscript on the Societies of Control." It is common to consider Deleuze's "Postscript" when discussing his relationship with the information, data, and algorithms that are now woven into the fabric of society. As a postscript, it seems that Deleuze slid one under the door—a message about the Internet society—at the end of his corpus, five years before his death and two years before *Critique et Clinique*. At the end of the book, the postscript seems like a note left at the corner of the desk as the place is shuttered and the lights turned off.

William Bogard compares control societies to panoptic spaces, following Deleuze's reading of Foucault's description of the visibility diagram that creates discreet spaces of entanglement for a number of individual bodies: school, barracks, prison, hospital. So, whereas a society that controls the body through enclosing spaces is structured around visibility (prisoners are seen without being able to see the guard in the central tower of the prison designed by Jeremy Bentham, and this visibility in turn can be seen in a classroom where all desks face the teacher who can scan the faces of the students), control is organized around "communication over distributed networks." Here, even though space is not the primary way of controlling the population, usernames and passwords do the trick:

> Instead of enclosing you, your body, they enclose your information. Your information does not flow serially between discrete spaces of control but is redistributed simultaneously and selectively across multiple networks, each protected by slightly modified codes, effecting a continuous modulation of control independent to location. "Visibility" does not organize these redistributions; codes do, in the sense that the passage of information within and between distributions entails having the right code.[23]

Following Hayles's perspective, we can see how the body is an afterthought in many of the works we have surveilled in this book. The art negotiates between information and perception. What becomes unnerving is the way that information can insidiously infiltrate the boundary of the body. What Deleuze suggests and what Bogard further fleshes out is what we had already encountered with the real self to data-self transformation of call center workers. Here, the transformation into a data-self does not have a pointed purpose like the call center workers, but is rather the established order to things: "Capital's project today is to engineer the disciplines directly into our DNA, which after all is just coded information. The final frontier in this project is to transform the socius into a distributed bio-network, whose relations nano-technologies can adjust in real time, all in the name of power and money."[24] We can already think of biostatistics that are compiled and uploaded onto corporate networks or unsolicited data poaching of electronic personal devices by manufacturers. Or simply of giving away copyright of personal images when uploading them to social networks. This data-self produced by the control society is named the dividual:

> Dividuals are database constructions, derived from rich, highly textured information on ranges of individuals that can be recombined in endless ways for whatever purposes. They are the abstract digital products of data-mining technologies and search engines and computer profiling, and they are the profiled digital targets of advertising, insurance schemes, and opinion polls. A dividual is a data distribution open to precise modulation, stripped down to whatever information construct is required for a specific intervention, task or transaction. Increasingly, postmodern subjectivity is defined by interaction with information meshes and the fractal dividuals they produce.[25]

Deleuze reads this notion of control already in William S. Burroughs's corpus as a new monster. He cites Foucault as having predicted it. And Virilio is a source of analysis of "ultrarapid forms of apparently free-floating control."[26] Confinement/open air surveillance is key and was already mentioned in *A Thousand Plateaus* in reference to Virilio: "Virilio concludes that the issue is less confinement than the management of the public ways, or the control of movement."[27] *A Thousand Plateaus* has already introduced several themes sharpened to the concise "Postscript" at the level of cybernetics and informational machines that form a "generalized regime of subjection," where this relationship is based on communication feedback between "third age" machines and humans.[28] The third age machine is taken up in the "Postscript": "Control societies function with a third generation of machines, with information technology and computers, where the passive danger is noise and the active, piracy and viral contamination."[29] Marshall McLuhan's theories articulate a similar relationship between machines and humans: "In machinic enslavement, there is nothing but transformations and exchanges of information, some of which are mechanical, others human."[30] The human/machine exchange is not based on the enclosure-analog-mold trifecta of discipline societies, but on the control-binary-modulation trifecta of control societies. Deleuze explains that if the enclosure societies operated according to molds that created distinct systems, modulation, or continuous changing, then deforming operations are at the heart of control society.[31] Modulating, changing, fluid-seeming exchange depends on code: passwords are what let you through thresholds set up in a control network.[32] Deleuze concludes by offering a couple of animals as emblems of societies past and present: the mole is the animal of the disciplinary society since it is the animal of enclosure, but the snake is the animal of the control society. Add to this list the drone (male bee after which the unmanned aircraft is named), surveilling from above as well as throughout the networks: released perhaps by the coding orchid. The snake slithers on the ground, but the drone flies above and works on the logic of mapping and surveilling.

I would like to propose three allegories to follow Galloway's invitation to refine a poetics of informational mapping. One by Brian Massumi, one by Günter Anders, and one by Marshall McLuhan. Massumi addresses the issue of cognitive mapping in relation to overcoding, revealing unintentional poetics at work in the mental representation of space. The poetics is at work in an individual's self-reflexive view of the process of mapping as well as in instances in which a map displays a lack of conscious control. The overwhelming amount of information one has to contend with in an instance of cognitive mapping overrides the proprioceptive system, over which one has no control. Massumi writes:

> No matter how consciously overcoding we like to be, our mappings are riddled with proprioceptive holes threatening at any moment to capsize the cognitive model (like the empty quarters filled with sea monsters on medieval maps). No matter how expert or encompassing our cognitive mapping gets, the monstrous sea of proprioceptive dead reckoning is more encompassing still. We are ever aswim in it.[33]

Massumi uses the language of the sea in order to give us a full picture of the chasm existing between our intention of mapping our environment and the fact that some of the information simply cannot be contained by our conscious mental apparatus. A true allegory of information and space: the cognitive map becomes the ship that will be capsized by the holes it contains. The ship then turns into a map, and the proprioceptive holes are the monsters on ancient maps where explorers have not finished translating reality into a representation. What this passage illustrates—folding back on Galloway's call for cognitive mapping amid a description of allegories of maps used in the process of capturing the known facets of our data reality—is the division between data and information, between what is accessible to us and what is not. Massumi's assessment raises issues of control in a closed-circuit networked society.

This notion of control is revealed in a story about a road, a king, and his son provided by Anders, a media theorist of the first half of the twentieth century who was interested in what he called promethean shame, or the inadequacy we feel in relation to machines. He writes:

> Because the king did not like how his son—straying from the roads under the king's control—was cutting across country in order to form his own image of the world, he gave him a horse and cart. "Now you no longer need to go on foot," were the king's words; "now you are no longer allowed to do so," was their meaning; "now you cannot do so anymore" was their result.[34]

The story can be read here in several ways. Anders creates a map drawn with predetermined outlets. The technological aspect that the son is dealing with as he loses more and more control due to his dependence on vehicles and roads is his relationship to the network, which becomes more and more circumscribed and therefore lost. The story could be illustrative of our overreliance on machines and how a certain level of comfort and convenience is gained as a proportional level of freedom of action and movement is lost. But what is really interesting is the way the king's son is caught in a network of roads. He can still see the fields he used to cut across but does not venture there. It serves the same function that Deleuze reads in his own projected future.

Deleuze describes in his Postscript a series of networks within a city and how an individual can make his way through them with the help of a code. The link with Anders's story about the king and his son is clear, but the situation is more extreme:

> Félix Guattari has imagined a town where anyone can leave their flat, their street, their neighborhood, using their (dividual) electronic card that opens this or that barrier, but the card may also be rejected on a particular day, or between certain times of day; it doesn't depend on the barrier but on the computer that is making sure everyone is in a permissible place, and effecting a modulation.[35]

The path the individual takes through the city is dependent on the network access card he or she possesses. Here the control is clear, not mere dependence on roads and vehicles like the king's son, but active tracking. Still the unrepresentable is part and parcel of this system: the code is not accessible; the paths are not chosen; but an impression of free range is given.

The embeddedness of the algorithm (or imperceptibility, as Parikka called it) is key to the control society. This notion, or a variation of it, was associated with media and was already understood by McLuhan when he provided a parable of his own, our third allegory. McLuhan ends *The Medium Is the Message* by tightening the knot around his media-being-extensions-of-human-senses thesis through tying in a social relationship string, stating that media "also configure the awareness and experience of each one of us."[36] He ends with the following observation taken from C.G. Jung, which reflects convincingly the king's roads parable advanced by Anders: "Every Roman was surrounded by slaves. The slave and his psychology flooded ancient Italy, and every Roman became inwardly, and of course unwittingly, a slave. Because living constantly in the atmosphere of slaves, he became infected through the unconscious with their psychology. No one can shield himself from such an influence."[37] Of course, replace "slave" with "media" and you solved the meaning behind the parable. Despite the historical cruelty of the analogy, the idea behind this substitution is one of feedback loops that create a cybernetic dependence between citizens caught in informational networks and helpful technologies. But again, this is not simply about the dependence on technology or media; it is about human obsolescence, to use Anders's word, and the overreliance on a network.

Elements of the three allegories—embeddedness (or the invisibility of the underpinning code); network control (or the fact that there is no freedom of movement beyond a coded path); and obsolescence (or a posthuman vision projected by functional images)—are present in Harun Farocki's films. At the turn of the millennium, Farocki's *Eye/Machine* trilogy (2001–03) provided a detached perspective on a surveillance society's new images: operative images [*Figure 20*]. All three allegories are in turn illustrated by operative images. *Eye/Machine I* starts with two silent identical moving images: grainy footage from above—or following Parks's phrase, "overhead images"—showing architectural structures on the ground through a framing of crosshairs that eventually explode. Silence is a defining characteristic of this type of image: "Another feature of surveillance footage is its silence, with CCTV in its early and most familiar forms either capturing only fuzzy grayscale images with no sound at all or images with a soundtrack that is too muffled or fragmented to be decoded by the human ear."[38] The dual image of grainy aerial landscapes with a graphic cross to calibrate the aim is slightly shifting toward featureless structures. The sequence ends with silent pale bursts that do not betray the on-ground damage from the missile strike this type of image captures. Farocki will couple these shots with surveillance footage of human figures walking slowly through parking lots and other public places. Other muted images abound: the dual picture of a man at café forgetting a briefcase that in turn blinks red; the aerial view of people coming through turnstiles, surrounded by pixelated lines [*Figure 21*]; pixelated arrows and lines superimposed on black-and-white image. These could illustrate the human figure in the surrounding environment explained by Massumi, or Guattari's ciphered city-path. Human figures appearing, heads on shoulders, seen from above are encircled in red, as if followed by a thin, progressing, amniotic fluid sac.

The juxtaposition between missiles and turnstiles has the camera equating war machine and human consumer. Later, a car camera focuses on the sidewalk, which is underlined in red. This could be an illustration of Anders's parable of the king's son, which also leads to network control since the son doesn't even need to be conscious of the path he is taking if he were to drive this car: "Many operational images show colored guidance lines, intended to portray the work of recognition. The lines tell us emphatically what is all important in these images, and just as emphatically what is of no importance at all."[39] The images explored in *Eye/Machine* are "not originally intended to be seen by humans but rather were supposed to function as an interface in the context of algorithmically controlled guidance processes. These images were meant to ensure the efficacy of a designated operation, which is why Farocki dubs them 'operational images.'"[40]

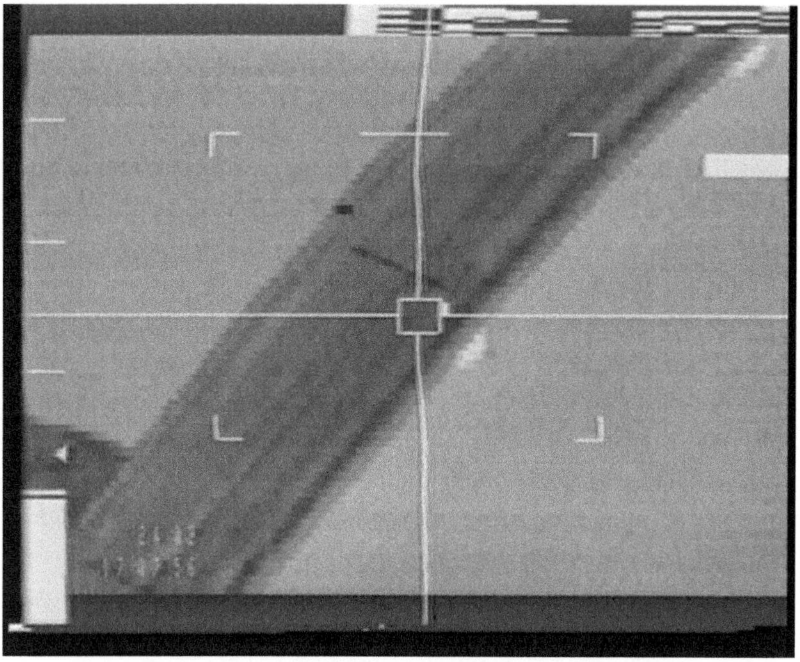

Figure 20 Harun Farocki, "Eye/Machine II." © Harun Farocki, 2002.

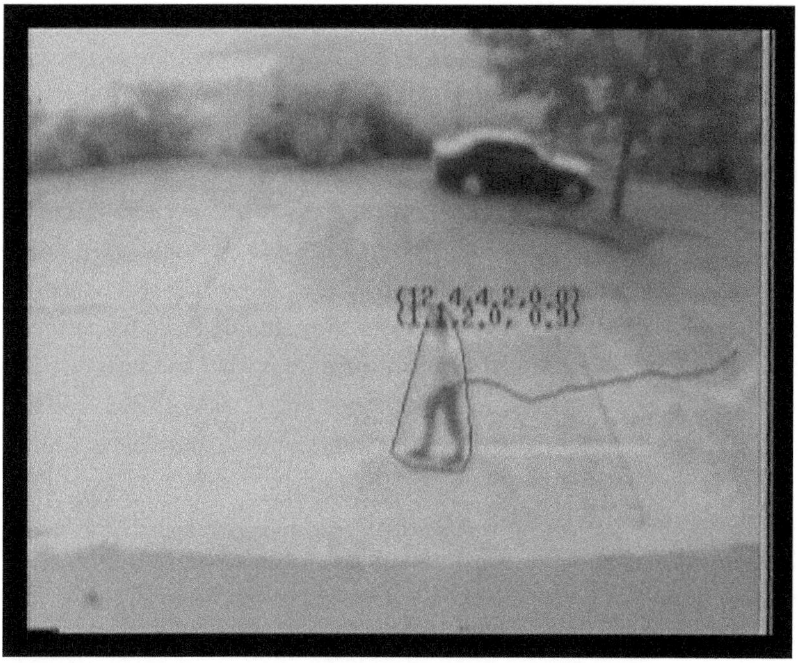

Figure 21 Harun Farocki, "Eye/Machine III." © Harun Farocki, 2003.

Väliaho links operational images to a politics: "Operational images take up the work of tools, more precisely, tools of power that impose a grid through which the world becomes visualized, intelligible, and, crucially, an object of manipulation."[41] The images that make up Farocki's films are "nonentertaining, noninstructive," and nonintentional. These images would exist even if they were never looked at by human eyes. They are all markers of an obsolescence of agency.[42]

Even though the film appears to show seemingly unrelated images, "what unites the many cases he discusses is the fading importance of the human as referential center in favor of 'intelligent machines' that render decisions autonomously."[43] These images operate purely by code. They are not concerned with spectatorship and barely acknowledge humans except as specs, measurements, statistics, biometrics: targets. The control society whose horizon Deleuze perceived is firmly entrenched.[44] In fact, the human factor is sidelined, not simply controlled, by this new type of image, as David Tomas explains. At the basis of this human exclusion is an image that no longer targets the human: "In this world, the concept of the *visible* image—the image produced for the human eye—has mutated."[45] This image, which can be seen by humans but, as Farocki's film demonstrates, has very questionable aesthetic qualities, is in fact a "by-product of a set of calculations which are the locus of the new immaterial numerical 'image.'"[46] The visual aspect of the image does not register aesthetically because it is not made for human senses. To become cognizant of the fact that the image is no longer targeting the human eye has an alienating consequence. The human's privileged place in relation to machines is on shaky grounds when "*placed* in situation where one not only becomes aware of one's precarious place in a world dominated by machines, but one also becomes vaguely conscious of a looming threat of historical redundancy."[47] The spatial *unheimlichkeit* yields to a temporal disorientation with the aesthetic image, effecting an unmooring of consciousness from place and history: "The video locates the viewer at the frontier of the posthuman algorithmic."[48] Tomas evokes a machinic landscape that dwarfs the human, but also sentient machines that progress without needing humans. The human is simply *de trop*: "The algorithm is an interface that denies both worlds simultaneously, and excludes all else as insubstantial because it is not anticipated in a program."[49] Farocki's operational image is a boundary marker that announces the great posthuman beyond:

> The operational image represents a mutation in the logic of data acquisition and management based on the development of a new relationship between worlds, as computer models increasingly "replace"—overlay—significant sections of concrete reality it also represents a significant augmentation in the penetrating powers of observation that can be measured through the proliferation of these models.[50]

What Farocki's images confront us with is something we had already touched on with Simon and Simondon: the lack of agency in the virtual element of the image: "*Eye/Machine* therefore functions as an introduction to a new kind of machine vision where there is no need of human intervention and therefore agency."[51] In fact, this dehumanizing draining of agency is demonstrated in the very aesthetics of the image presented by Farocki, as explained by Martin Blumenthal-Barby, who sees in the colorfully pixelated lines struggling for stylistic fluidity atop grainy clumsily framed black-and-white images a visual reference to an Action painting–type of abstraction. These images fall under the purview of the aesthetic category of "uncalculated beauty": "We encounter this 'uncalculated beauty' (and its invocation of the 'optical unconscious') recurrently in Eye/Machine in the form of operational images, which in their bright blue, red, green, and yellow color patterns echo the abstract works of contemporary digital artists."[52] The "optical unconscious," discovered by Benjamin at the dawn of the photographic image, is now surpassed by the US military, as Farocki reminds us.[53] These are so far removed from what early photographs revealed of the world in their immovable capture of quick daily details that they escape the grasp of aesthetic sense and fall into a sort of rudderless abstraction. The color maps out special elements in the pictures; and the operative system attached to the camera in each case needs to detect these elements in order to react. There is no space for contemplation.

Conclusion

I will end with this image, which in turn can be seen as an allegory of the posthuman operative image. One of the most pathos-imbued images in Farocki's film is that of a man slowly walking beside his bed in what could be a retirement home. The image is grainy, but you can see the bed and a table with chairs making up part of a sitting area. The gray, ill-defined silhouette shuffles from left to right, framed within an unsteady vertical green rectangle. At one point, the man unceremoniously falls. The surrounding frame does too. Now it is a horizontal rectangle. No one comes in to help. The rectangle blinks red.

Notes

1 Jakub Zdebik, "Skin Aesthetics as Incarnation: Gilles Deleuze's Diagram of Francis Bacon," *ESC* 34, no. 1 (March 2008): 153.
2 Vytas Narusevicius, "Walid Raad's Double Bind: The Atlas Group Project, 1989–2004," *RACAR* 39, no. 2 (2014): 43.
3 Ibid.
4 Hajra Waheed's Artist's website: http://hajrawaheed.com/
5 Ibid.
6 Ibid.
7 Lisa Parks, "Zeroing In: Overhead Imagery, Infrastructure ruins, and Datalands in Afghanistan and Iraq," in *The Visual Culture Reader*, ed. Nicholas Mirzoeff (London: Routledge, 2013), 197.
8 Ibid.
9 Ibid.
10 Ibid.
11 Michel Foucault, *The Order of Things* (London: Routledge, 2005), 422.
12 Krauss, "Grids," 50.
13 N. Katherine Hayles, *How We Became Posthuman: Virtual Bodies in Cybernetics, Literature and Informatics* (Chicago: Chicago University Press, 1999), 2.
14 Ibid., 3.
15 Ibid.
16 Bruno Latour, *We Have Never Been Modern*, trans. Catherine Porter (Cambridge: Harvard University Press, 1993), 136.
17 Ibid., 138.
18 Galloway, *The Interface Effect*, 98. Compare to: "Perhaps 'digital' as an ontology of binary coding into 0s and 1s is not a regime of creative differences, but at the same time the 'digital' is too broad a category for any specific understanding of the weird materialities of network culture; software is increasingly more about relationality within its code world but also in its relations to the outsides in which it is embedded (from abstract machines of capitalism to covert assemblages such as games, browsers or, for example, mobile phone interfaces)." See Parikka, "Ethologies of Software Art," 119.
19 Galloway, *The Interface Effect*, 98.

20 Ibid.
21 Galloway, *The Interface Effect*, 99.
22 Ibid.
23 William Bogard, "Deleuze and Machines: A Politics of Technology?," in *Deleuze and New Technology*, ed. Mark Poster and David Savat (Edinburgh: Edinburgh University Press, 2009), 21.
24 Ibid.
25 Ibid.
26 Deleuze, *Negotiations*, 178.
27 Deleuze and Guattari, *Thousand Plateaus*, 558n63.
28 Ibid., 458.
29 Deleuze, *Negotiations*, 180.
30 Deleuze and Guattari, *Thousand Plateaus*, 458.
31 Deleuze, *Negotiations*, 178–79.
32 Ibid., 180.
33 Brian Massumi, "Strange Horizon: Buildings, Biograms and the Body Topologic," *Parables for the Virtual* (Durham: Duke University Press, 2002), 181.
34 Christopher John Müller, *Prometheanism: Technology, Digital Culture and Human Obsolescence* (London: Rowman & Littlefield International, 2016), 129.
35 Deleuze, *Negotiations*, 181–82.
36 Marshall McLuhan, "The Medium Is the Message," in *Understanding Media: The Extensions of Man* (New York: McGraw-Hill, 1964), 35.
37 Ibid.
38 Annie Ring, "System Error: Complicity with Surveillance in Contemporary Workplace Documentaries," *Seminar: A Journal of Germanic Studies* 52, no. 4 (2016): 486, available online: http://www.utpjournals.press/doi/abs/10.3138/seminar.52.4.07 (accessed June 29, 2017).
39 Harun Farocki's website: http://www.harunfarocki.de/home.html
40 Martin Blumenthal-Barby, "'Cinematography of Devices': Harun Farocki's Eye/Machine Trilogy," *German Studies Review* 38, no. 2 (2015): 329, available online: https://scholarship.rice.edu/bitstream/handle/1911/81875/38.2.blumenthal-barby.pdf?sequence=1 (accessed February 13, 2017).
41 Väliaho, *Biopolitical Screen*, 64.
42 A more tempered view of the human experience of new technologies in the twenty-first century would be in Mark Hansen's reading of Whitehead as a media theorist and the way that the humans interact with technologies that are imperceptible yet instrumental to the molding of experience: "Human experience is currently undergoing a fundamental transformation caused by the complex entanglement of humans within networks of media technologies that operate predominantly, if not almost entirely, outside the scope of human modes of

awareness (consciousness, attention, sense perception, etc.)." Mark B. N. Hansen, *Feed-Forward: On the Future of Twenty-First-Century Media* (Chicago: University of Chicago Press, 2015), 5. Folded within an environment enmeshed by computer networks, sensing technologies and even smartphones, human experience is redefined by what is not perceived: "Far more of what goes on in our daily lives is carried out by machines functioning at their own timescales, meaning outside of our direct perceptual grasp but in ways that do significantly affect our activity." Ibid., 23.

43 Blumenthal-Barby, "Cinematography," 330.

44 Ibid., 333.

45 David Tomas, *Vertov, Snow, Farocki: Machine Vision and the Posthuman* (New York: Bloomsbury, 2013), 235.

46 Ibid.

47 Ibid.

48 Ibid.

49 Ibid.

50 Ibid.

51 Ibid.

52 Blumenthal-Barby, "Cinematography," 333.

53 Ibid.

Conclusion: Tracing on the Map

What territory does the map-image cover? What does a map drawn in the previous pages look like when it is unfolded and seen at one glance? Borges's map, for example, covered a whole empire. That map was hard to imagine, its image flickering inconsistently in the imagination as we tried to attain an ensemble view of its mechanics. Because of this, it is the de facto emblem for the concept of the map-image and its fluctuating quality. We have seen the distortion the map undergoes depending on which medium in Deleuze's (and Guattari's) corpus it is applied to. In painting, a geographical landscape emerges: desert, marshland, and ocean that are gray, sloppy, and split heads. It is a haptic map that should not be too tactile or it would scramble the surface of the painting, like in abstract art. In cinema, maps are superimposed, they modulate between time and space, and they orient the screen toward electronic images. And in literature, maps of spaces such as castles, hotels, and burrows with their multiple corridors, entrances, and tunnels morph into thresholds of intensity and cease to be spatial altogether.

Even a more conventional map, a representation of a geographical map, like the one hanging in the background of a Vermeer painting, is unstable. Not only because baroque map-makers had not yet settled on a conventional orientation of the map itself, with north being on top and south at the bottom and so forth, but because the map behind the figures either self-reflexively traces a pathway back to the surface of the painting (centripetal motion) or suggests a line of flight out of the frame (centrifugal motion). Vermeer's map also traveled through time and through media: from painting to photography to cinema, always keeping its play of light, shadow, and color true to the source, as well as the tracing back to the original equivocation of the map with the blank screen. The shifty nature of the map finds its source in Steinberg's flatbed picture plane, functioning like a pivot point between nature and culture, between the vertical and the horizontal, and between illusion and information. This device is taken up by Deleuze to theorize the digital from painting, as a window to the opaque grid screen of information.

The map is rhizomatic: it is open, it connects, it is removable and reversible. It is in constant flux. It is art; it can be hung on a wall. It is abstract: a site of multiplicities and diagrammatic modifications. But by being rhizomatic, it is also a copy, a trace, a redundancy, and a code. With the code, the map does not necessarily abandon its dynamism. In fact, it is through coding that it serves as a transductive device between milieus. But the code is stochastic, dependent on chance, random but transitional. And this code, in effect, links multiple disciplines from entomology to biology to psychology to aesthetics. The aesthetics of code relies on mapping and in particular on the map-image since visualization of information negotiates between imagined abstract geographies: network cartographies of unrepresentable spaces. And these spaces are manifest in painting as well as in algorithmic art in the formalization of data, whether analog or digital.

But a map-image can be a disembodied cartography of boundaries, strata, and lines. Functioning through the process of translation between media, geographical locations, or theoretical milieus. It is a geography of the surface of art, the rift in language, and the zone between striations (theoretical and representational). It is an abstract cartography of global economic conditions, of urban territoriality, and of the blank screen. It is the map-image coding information onto art. It is an oscillating, vibratory map of the in-between.

The map-image negotiates between the virtual and the actual. Consequently, it is instrumental in the analysis of abstract spaces represented in digital art. These spaces can be fluctuating masses of discreet elements such as swarms; or abstract, constantly evolving digital environments; or even cityscapes that are part painting, part map, part diagram, and part pixelated grid. Here the mapped-out abstract environments manifest themselves in the form of planes consisting of infinite folds of the virtual unfolding onto the actual: a cosmos of virtual particles, inorganic folds resembling teeming fish hatcheries, systems of swarming particles, nonhuman-scaled algorithmic landscapes, intra-perceptive images imbedded into environments about to be triggered into actuality, microstructural, geometric, and contrasting motifs of decorative surfaces, and grids—spatial, organizational, and artistic grids.

At other times the map-image is an island. Or images of various islands: island as Internet art, as game space sculpture, as plot-propelling device of a science fiction film. The island is an image of isolation, framing, and preservation. The island as image becomes a fluctuating spatial critical device of cartographic art. The island as map-image is a condition for creativity but a marker of potential obliteration. The island is the map-image of artistic creation and dystopic desertification.

The map-image captures overhead images (satellite, ballistic, and spy cartographies) as well as operative images (machine-made images not destined for human eyes)—new dimensions of posthuman experiences. By charting these new vistas of surveillance and control, the map-image is a warning device of things to come. As a cartography of informational processes, the map-image reveals the invisible underpinnings of code in a surveillance society, the control of network that offers no possibility of freedom beyond a predetermined path and the obsolescence of human visual aesthetics.

Throughout these pages, maps, the cartographic elements, and geographical images formed a network, an assemblage of disparate, multiple parts that, depending on their immediate formation, took on particular functions. This peripatetic organization of terms surrounding Deleuzean cartography created a concept that constantly changed shapes while remaining delimited. Notions surrounding networks, assemblages and *agencements*, forming the matrix for the map-image as a critical device for visual art, need to be explored next in order to deepen the relationship between art history and Deleuze's philosophical system.

Works Cited

Agamben, Giorgio. *What Is an Apparatus? and Other Essays*. Translated by David Kishik and Stefan Pedatella. Stanford: Stanford University Press, 2009.

Alpers, Svetlana. *The Art of Describing: Dutch Art in the Seventeenth Century*. Chicago: University of Chicago Press, 1983.

Arcangel, Cory. "Recent Experiments in Modern Composition, Software, and Stand-Up Comedy." The Art, Technology and Culture Colloquium. UC Berkeley's Centre for New Media. Audio recording. Available online: http://atc.berkeley.edu/lectures/mp3s/Cory%20Arcangel.mp3 (accessed June 29, 2017).

Arcangel, Cory. "Untitled Translation Exercise." Available online: http://www.coryarcangel.com/things-i-made/2006-002-untitled-translation-exercise (accessed April 12, 2016).

Astoff, Jens, "Janice Kerbel." *Artforum*, July 15, 2004. Available online: http://artforum.com/archive

Augé, Marc. *Non-Lieux: Introduction à une anthropologie de la surmodernité*. Paris: Éditions du Seuil, 1992.

Barthélémy, Jean-Hugues. "Glossary: Fifty Key Terms in the Works of Simondon." In *Gilbert Simondon: Being and Technology*, edited by Arne De Beover et al., translated by Arne De Boever, 203–31. Edinburgh: Edinburgh University Press, 2012.

Beckman, Karen. "Animating the *Cinéfils*: Alain Resnais and the Cinema of Discovery." *Cinema Journal* 54, no. 4 (Summer 2015): 1–25.

Benjamin, Walter. "The Task of the Translator: An Introduction to the Translation of Baudelaire's Tableaux parisiens." In *Illuminations: Essays and Reflections*, edited by Hannah Arendt, translated by Harry Zohn, 69–83. New York: Schocken Books, 2007.

Benjamin, Walter. "The Work of Art in the Age of Mechanical Reproduction." In *Illuminations: Essays and Reflections*, edited by Hannah Arendt, translated by Harry Zohn, 217–51. New York: Schocken Books, 1969.

Berkebile McManus, Elizabeth. "Protecting the Island: Narrative Continuance in *Lost*." *Journal of the Fantastic in the Arts* 22, no. 1 (81) (2011): 4–23.

Bertens, Hans. *Literary Theory: The Basics*. London: Routledge, 2014.

Biederman, Charles. *Art as the Evolution of Visual Knowledge*. Red Wing: Charles Biederman, 1948.

Biederman, Charles. "The Real and the Mystic in Art and Science." In *Theories and Documents of Contemporary Art: A Sourcebook of Artists' Writings*, edited by Kristine Stiles and Peter Selz, 81–82. Berkeley: University of California Press, 1996.

Birnbaum, Dara, and Cory Arcangel. "Do It 2." *Artforum International* 7 (2009): 190–98.

Blanchot, Maurice. *The Infinite Conversation*. Translation and foreword by Susan Hanson. Minneapolis: University of Minnesota Press, 1993.

Blumenthal-Barby, Martin. "'Cinematography of Devices': Harun Farocki's Eye/Machine Trilogy." *German Studies Review* 38, no. 2 (2015): 329–51. Available online: https://scholarship.rice.edu/bitstream/handle/1911/81875/38.2.blumenthal-barby.pdf?sequence=1 (accessed February 13, 2017).

Bogard, William. "Deleuze and Machines: A Politics of Technology?" In *Deleuze and New Technology*, edited by Mark Poster and David Savat, 15–31. Edinburgh: Edinburgh University Press, 2009.

Bogue, Ronald. *Deleuze on Literature*. New York: Routledge, 2003.

Bois, Yve-Alain, and Amy Reiter-McIntosh. "Piet Mondrian, 'New York City.'" *Critical Inquiry* 14, no. 2 (1988): 244–77.

Borges, Jorge Luis. "On Exactitude and Science." In *The Aleph*. London: Penguin, 2000.

Brito, Vanessa. "L'île déserte et le peuple qui manque." In *Gilles Deleuze et les images*, edited by François Dosse and Jean-Michel Frodon, 64–74. Paris: Cahiers du Cinéma Institut National de l'Audiovisuel, 2008.

Buci-Glucksmann, Christine. "Of the Diagram in Art." Translated by Josh Wise. *ANY: Architecture New York* 23 (1998): 34–36.

Buci-Glucksmann, Christine. "Temps et modèles du virtuel." In *L'art à l'époque du virtuel*, edited by C. Buci-Glucksmann, 9–24. Paris: L'Harmattan, 2003.

Burgoyne, Robert. "Customising Pleasure: 'Super Mario Clouds' and the John Ford Sky." In *Cinephilia in the Age of Digital Reproduction*, edited by Scott Balcerzak and Jason Sperb, 160–70. London: Wallflower Press, 2009.

Cohen, Tom. "Toxic Assets: de Man's Remains and the Ecocatastrophic Imaginary (an American Fable)." In *Theory and the Disappearing Future*, edited by Tom Cohen, Claire Colebrook and J. Hillis Miller, 89–129. London: Routledge, 2012.

Conley, Tom. *Cartographic Cinema*. Minneapolis: University of Minnesota Press, 2007.

Conley, Tom. "The Desert Island." In *Deleuze and Space*, edited by Ian Buchanan and Gregg Lambert, 207–19. Edinburgh: Edinburgh University Press, 2005.

Conley Tom. "The Strategist and the Stratigrapher." In *Afterimages of Gilles Deleuze's Film Philosophy*, edited by D. N. Rodowick, 193–21. Minneapolis: University of Minnesota Press, 2010.

Craft, Catherine. *An Audience of Artists: Dada, Neo-Dada, and the Emergence of Abstract Expressionism*. Chicago: University of Chicago Press, 2012.

Culp, Andrew. *Dark Deleuze*. Minneapolis: University of Minnesota Press, 2016.

Cunnally, John. "Antiquarianism and the Origins of the Flatbed Matrix." *Notes in the History of Art* 31/32, no. 4/1 (2012): 6–12.

Daly, Kristen. "Cinema 3.0: The Interactive-Image." *Cinema Journal* 50, no. 1 (2010): 81–98.

Damisch, Hubert. "Genealogy of the Grid." In *The Archive of Development*, edited by Henk Slager and Annette W. Balkema. Amsterdam: Rodopi, 1998.
Danto, Arthur. "The Artworld." *The Journal of Philosophy* 19 (1964): 571–84.
de Man, Paul. "Conclusions: Walter Benjamin's 'The Task of the Translator.'" In *The Resistance to Theory*, 73–105. Minneapolis: University of Minnesota Press, 1986.
de Singly, Camille. *Guido Molinari: peintre moderniste canadien: Les espaces de la carrière*. Preface by François-Marc Gagnon. Paris: L'Harmattan, 2004.
Debord, Guy. *Society of the Spectacle*. Detroit: Black & Red, 1983.
Deleuze, Gilles. "The Actual and the Virtual." In *Dialogues II*, translated by Eliot Ross Albert, 148–59. New York: Columbia University Press, 2002.
Deleuze, Gilles. "Causes et raisons des îles désertes." In *le Déserte et autres textes: Textes et entretiens 1953–1974*, edited by David Lapoujade, 11–17. Paris: Minuit, 2002.
Deleuze, Gilles. "Causes and Reasons of Desert Islands." In *Desert Islands and Other Texts: 1953–1974*, edited by David Lapoujade, translated by Michael Taormina, 9–14. Los Angeles: Semiotext(e), 2002.
Deleuze, Gilles. *Cinema 2: The Time-Image*. Translated by Hugh Tomlinson and Robert Galeta. Minneapolis: University of Minnesota Press, 1997.
Deleuze, Gilles. *Difference and Repetition*. Translated by Paul Patton. New York: Columbia University Press, 1994.
Deleuze, Gilles. *Essays Critical and Clinical*. Translated by Daniel W. Smith and Michael A. Greco. Minneapolis: University of Minnesota Press, 1997.
Deleuze, Gilles. *The Fold: Leibniz and the Baroque*. Translated by Tom Conley. London: Athlone, 1993.
Deleuze, Gilles. *Francis Bacon: The Logic of Sensation*. Translated by Daniel F. Smith. London: Continuum, 2003.
Deleuze, Gilles. *Logic of Sense*. Edited by Constantin V. Boundas, translated by Mark Lester. New York: Columbia University Press, 1990.
Deleuze, Gilles. *Negotiations*. Translated by Martin Joughin. New York: Columbia University Press, 1995.
Deleuze, Gilles. "What Children Say." In *Essays Critical and Clinical*, translated by Daniel W. Smith and Michael A. Greco, 61–67. Minneapolis: University of Minnesota Press, 1997.
Deleuze, Gilles, and Félix Guattari. *Anti-Oedipus: Capitalism and Schizophrenia*. Translated by Robert Hurley, Mark Seem, and Helen R. Lane. Minneapolis: University of Minnesota Press, 2000.
Deleuze, Gilles, and Félix Guattari. *Kafka: Toward a Minor Literature*. Translated by Dana Polan. Minneapolis: University of Minnesota Press, 2003.
Deleuze, Gilles, and Félix Guattari. *A Thousand Plateaus: Capitalism and Schizophrenia II*. Translated by Brian Massumi. Minneapolis: University of Minnesota Press, 2005.
Deleuze, Gilles, and Félix Guattari. *What Is Philosophy?* Translated by Hugh Tomlinson and Graham Burchell. New York: Columbia University Press, 1994.

Dhillon, Kim. "See It Now or Miss It Forever: Materiality, Visuality, and the Written Word in Janice Kerbel's Recent Artwork." *RACAR* 36, no. 1 (2011): 24.
Dick, Terence. "The Mapmaker." *Canadian Art* 23, no. 4 (2006): 50.
Eliasson, Olafur, and Aaron Betsky. "Let It Rain: The Art of the Code." In *Studio Olafur Eliasson: An Encyclopedia*, edited by Anna Engberg-Pedersen. Köln: Tachen, 2008.
Elkins, James. "Art History and Images That Are Not Art." In *Images: Critical and Primary Sources: Understanding Images*, edited by Sunil Manghani, 554–71. London: Bloomsbury, 2013.
Elson, Linda G. *Paradox Lost: A Cross-Contextual Definition of Levels of Abstraction.* Cresskill: Hampton Press, 2010.
Evens, Aden L. "Actual/Virtual." In *Understanding Deleuze, Understanding Modernism*, edited by Paul Ardoin et al., 247–48. New York: Bloomsbury, 2014.
Feijis, Loe. "Divisions of the Plane by Computer: Another Way of Looking at Mondrians' Nonfigurative Compositions." *Leonardo* 37, no. 3 (2004): 217–22.
Ferris, David S. *The Cambridge Introduction to Walter Benjamin*. Cambridge: Cambridge University Press, 2008.
Flaxman, Gregory. *Gilles Deleuze and the Fabulation of Philosophy*. Minneapolis: University of Minnesota Press, 2012.
Foucault, Michel. "Of Other Spaces." Translated by Jay Miskowiec. *Diacritics* 16, no. 1 (1986): 22–27.
Foucault, Michel. *The Order of Things*. London: Routledge, 2005.
Galloway, Alexander R. *The Interface Effect*. Cambridge: Polity Press, 2012.
Galloway, Alexander R. "*StarCraft*, or, Balance." *Grey Room* 28 (2007): 86–109.
Godfrey, Mark. "First Take: Mark Godfrey on Janice Kerbel." *Artforum* 44, no. 5 (2006): 183.
Greenberg, Clement. "Towards a Newer Laocoon." In *Art in Theory: 1900–1990: An Anthology of Changing Ideas*, edited by Charles Harrison and Paul Wood, 554-60. Oxford: Blackwell, 1996.
Grosz, Elizabeth. *Architecture from the Outside: Essays on Virtual and Real Space.* Cambridge: MIT Press, 2001.
Hallward, Peter. *Out of This World: Deleuze and the Philosophy of Creation*. London: Verso, 2006.
Hansen, Mark B. N. *Feed-Forward: On the Future of Twenty-First-Century Media.* Chicago: University of Chicago Press, 2015.
Hayes, Brian. "First Links in the Markov Chain." *American Scientist* 101 (2013): 92–97.
Hayles, N. Katherine. *How We Became Posthuman: Virtual Bodies in Cybernetics, Literature and Informatics*. Chicago: Chicago University Press, 1999.
Heidegger, Martin. "The Origin of the Work of Art." In *Poetry, Language, Thought*, translated by Albert Hofstadter, 15–87. New York: Perennial Classic, 2001.
Heidegger, Martin. "The Thing." In *Poetry, Language, Thought*, translated by Albert Hofstadter, 161–84. New York: Perennial Classic, 2001.

Heinze, Rüdiger. "'This Makes No Sense at All': Heterarchy in Fictional Universes." *Story Worlds: A Journal of Narrative Studies* 7, no. 2 (2015): 76–91.

Hölling, Hanna B. *Revisions—Zen for Film*. New York: Bard Graduate Center, 2015.

Hölzel, Eduard. "Holographic Map." In *The History of Cartography, vol. 6, Cartography in the Twentieth Century. Part 1*, edited by Mark Monmonier, 619–20. Chicago: University of Chicago Press, 2015.

Huyssen, Andreas. "Guillermo Kuitca: Painter of Space." In *Guillermo Kuitca Everything: Paintings and Works on Paper, 1980–2008*, edited by Douglas Dreishpoon, 22–33. London: Scala, 2009.

Iversen, Margaret. "Index, Diagram, Graphic Trace." *Tate Papers* 18 (2012). Available online: http://www.tate.org.uk/research/publications/tate-papers/18/index-diagram-graphic-trace (accessed May 5, 2017).

Jacob, Christian J., Gerald Hushlak, Jeffrey E. Boyd, Paul Nuytten, Maxwell Sayles, and Marcin Pilat. "'SwarmArt': Interactive Art from Swarm Intelligence." *Leonardo* 40, no. 3 (2007): 248–55.

Jameson, Fredric. *The Prison-House of Language: A Critical Account of Structuralism and Russian Formalism*. Princeton: Princeton University Press, 1974.

Jameson, Fredric. "Utopianism after the End of Utopia." In *Postmodernism, or, The Cultural Logic of Late Capitalism*, 154–80. Durham: Duke University Press, 1991.

Joselit, David. "On Aggregators." *October* 146 (2013): 3–18.

Kane, Carolyn L. *Chromatic Algorithms: Synthetic Color, Computer Art, and Aesthetics after Code*. Chicago: University of Chicago Press, 2014.

Kane, Carolyn L. "Error as Critical Praxis." *Millennium Film Journal*, no. 56 (2012): 7–8.

Kaplan, Louis. *László Moholy-Nagy: Biographical Writings*. Amherst: Duke University Press, 1995.

Kearns, Chris. "Rereading Borges Rereading: A Question of Marginality." *Symplokē* 1, no. 2 (1993): 121–39.

Kerslake, Christian. *Deleuze and the Unconscious*. London: Continuum, 2007.

Kolb, William M. "Blade Runner Film Notes." In *Retrofitting Blade Runner: Issues in Ridley Scott's Balde Runner and Philip K. Dick's Do Androids Dream of Electric Sheep?* edited by Judith B. Kerman, 154–77. Bowling Green: Bowling Green State University Popular Press, 1991.

Krapp, Peter. *Noise Channels: Glitch and Error in Digital Culture*. Minneapolis: University of Minnesota Press, 2011.

Krauss, Rosalind. "Grids." *October* 9 (1979): 50–64.

Krauss, Rosalind. "The Slung Leg Hypothesis." *October* 136 (2011): 218–21.

Krauss, Rosalind. "Stieglitz/'Equivalents.'" *October* 11 (1979): 129–40.

Latour, Bruno. "On Actor-Network Theory: A Few Clarifications." *Soziale Welt* 47, no. 4 (1996): 369–81.

Latour, Bruno. *We Have Never Been Modern*. Translated by Catherine Porter. Cambridge: Harvard University Press, 1993.

Lévy, Pierre. *Qu'est-ce que le virtuel?* Paris: Éditions La Découverte, 1995.
Lingis, Alphonso. "Deleuze on a Deserted Island." In *Philosophy and Non-philosophy since Merleau-Ponty*, edited by Hugh Silverman, 152–74. Evanston: Northwestern University Press, 1997.
Lisson Gallery. "Lisson Presents 7." Available online: http://www.lissongallery.com/exhibitions/lisson-presents-7 (accessed April 13, 2016).
Lovatt, Anna. "Janice Kerbel: Diagramming Desire." *Afterall: A Journal of Art, Context and Enquiry* 37 (Autumn/Winter 2014): 90–101.
Luhmann, Niklas. *Art as a Social System*. Translated by Eva M. Knodt. Stanford: Stanford University Press, 2000.
Lyotard, Jean-François. *Discours, Figure*. Paris: Klincksieck, 1971.
Lyotard, Jean-Francois. "What Is Postmodernism?" In *Art in Theory 1900–1990: An Anthology of Changing Ideas*, edited by Charles Harrison and Paul Wood, 1008–15. Oxford: Blackwell, 1996.
Marks, Laura U. *Enfoldment and Infinity: An Islamic Genealogy of New Media Art*. Cambridge: The MIT Press, 2010.
Marks, Laura U. "Infinity and Accident: Strategies of Enfoldment in Islamic and Computer Art." *Leonardo* 39, no. 1 (2006): 37–42.
Martin, Jean-Clet. *Bréviaire de l'éternité: Vermeer et Spinoza*. Paris: Éditions Léo Scheer, 2011.
Martin, Jean-Clet. "Le virtuel incarné." In *L'art à l'époque du virtuel*, edited by Buci-Glucksmann, 197–210. Paris: L'Harmattan, 2003.
Massumi, Brian. "The Autonomy of Affect." In *Parables for the Virtual: Movement, Affect, Sensation*, 23–45. Durham: Duke University Press, 2002.
Massumi, Brian. "Strange Horizon: Buildings, Biograms and the Body Topologic." In *Parables for the Virtual*, 177–207. Durham: Duke University Press, 2002.
McLuhan, Marshall. "The Medium Is the Message." In *Understanding Media: The Extensions of Man*. New York: McGraw-Hill, 1964.
Meltzer, Eve. *Systems We Have Loved: Conceptual Art, Affect, and the Antihumanist Turn*. Chicago: University of Chicago Press, 2013.
Mesei, Leslie. "Randomness in Computer Graphics." In *Cybernetics, Art and Ideas*, edited by Jasia Reichardt, 165–74. Greenwich: New York Graphic Society Ltd., 1971.
Miller, J. Hillis. "Paul de Man at Work: In These Bad Days, What Good Is an Archive?" In *Theory and the Disappearing Future: On de Man, On Benjamin*, edited by Tom Cohen, Claire Colebrook, and J. Hillis Miller, 55–88. London: Routledge, 2012.
Milroy, Sarah. "Francine Savard's Weather-Forecast Works Shine Bright." *Canadian Art*, 4 November 2013. Available online: http://www.diazcontemporary.ca/Artists_SavardReviews.html
More, Thomas. "Utopia." In *Three Early Modern Utopias: Utopia, New Atlantis, The Isle of Pines*, edited by Susan Bruce, 1–148. Oxford: Oxford University Press, 1999.

Morton, Timothy. "Ecology as Text, Text as Ecology." *The Oxford Literary Review* 32, no. 1 (2010): 1–17.

Morton, Timothy. *Ecology without Nature: Rethinking Environmental Aesthetics*. Cambridge: Harvard University Press, 2007.

Mulhall, Stephen. "Zhora through the Looking-Glass: Notes on an Esper Analysis of Leon's Photograph." In *Blade Runner*, edited by Amy Coplan and David Davies. London: Routledge, 2015.

Mullarkey, John. *Post-Continental Philosophy: An Outline*. London: Continuum Press, 2006.

Müller, Christopher John. *Prometheanism: Technology, Digital Culture and Human Obsolescence*. London: Rowman & Littlefield International, 2016.

Narusevicius, Vytas. "Walid Raad's Double Bind: The Atlas Group Project, 1989–2004." *RACAR* 39, no. 2 (2014): 43–53.

Owens, Craig. "The Allegorical Impulse: Toward a Theory of Postmodernism." *October* 12 (1980): 67–86.

Parikka, Jussi. "Ethologies of Software Art: What Can a Digital Body of Code Do?" In *Deleuze and Contemporary Art*, edited by Stephen Zepke and Simon O'Sullivan, 117–33. Edinburgh: Edinburgh University Press, 2010.

Parks, Lisa. "Zeroing In: Overhead Imagery, Infrastructure Ruins, and Datalands in Afghanistan and Iraq." In *The Visual Culture Reader*, edited by Nicholas Mirzoeff, 196–206. London: Routledge, 2013.

Pegg, Danny. "Prometheus." *Journal of Religion & Film* 17, no. 2 (2013): 6–7. Available online: http://digitalcommons.unomaha.edu/jrf/vol17/iss2/17

Pethö, Ágnes. *Cinema and Intermediality: The Passion for the In-Between*. Newcastle upon Tyne: Cambridge Scholars Publishing, 2011.

Piaget, Jean, and Bärbel Inhelder. *La représentation de l'espace chez l'enfant*. Paris: PUF, 1947.

Rajewsky, Irina P. "Border Talks: The Problematic Status of Media Borders in the Current Debate about Intermediality." In *Media Borders, Multimodality and Intermediality*, edited by Lars Elleström, 51–68. New York: Palgrave Macmillan, 2010.

Raley, Rita. "Statistical Material: Globalization and the Digital Art of John Klima." *The New Centennial Review* 3, no. 2 (2003): 67–89.

Rancière, Jacques. "Are Some Things Unrepresentable?" In *The Future of the Image*, translated by Gregory Elliott, 109–38. New York: Verso, 2007.

Ring, Annie. "System Error: Complicity with Surveillance in Contemporary Workplace Documentaries." *Seminar: A Journal of Germanic Studies* 52, no. 4 (2016): 456–89. Available online: https://doi.org/10.3138/seminar.52.4.07 (accessed February 13, 2017).

Ropars-Wuilleumier, Marie-Claire. "Image or Time? The Thought of the Outside in *The Time-Image* (Deleuze and Blanchot)." In *Afterimages of Gilles Deleuze's Film*

Philosophy, edited by D. N. Rodowick, 15–30. Minneapolis: University of Minnesota Press, 2010.

Rose-Redwood, Reuben S. "Genealogies of the Grid: Revisiting Stanislawski's Search for the Origin of the Grid-Pattern Town." *Geographical Review* 98, no. 1 (2008): 42–58.

Rosenstiehl, Pierre, and Jean Petitot. "Automate asocial et systèmes acentrés." *Communications* 22, no. 1 (1974): 45–62. doi: 10.3406/comm.1974.1337

Ross, Miriam. "Stereoscopic Visuality: Where Is the Screen, Where Is the Film?" *Convergence: The International Journal of Research into New Media Technologies* 19, no. 4 (2013): 406–14.

Ruyer, Raymond. *La genèse des formes vivantes*. Paris: Flammarion, 1958.

Sammon, Paul M. *Future Noir: The Making of Blade Runner*. New York: HarperCollins, 1996.

Sauvagnargues, Anne. *Deleuze et l'art*. Paris: PUF, 2005.

Sauvagnargues, Anne. "Desiring Machines and Social Coding." In *Artmachines: Deleuze, Guattari, Simondon*, translated by Suzanne Verderber with Eugene W. Holland, 195–218. Edinburgh: Edinburgh University Press, 2016.

Shone, Richard. "Rachel Whiteread's 'House.'" *The Burlington Magazine* 135, no. 1089 (1993): 837–38.

Simon Jr., John F. Artist's website. http://www.numeral.com

Simondon, Gilbert. *Imagination et invention (1965–66)*. Paris: Éditions de la transparence, 2008.

Sloterdijk, Peter. *Sphères III: Écumes: Sphérologie plurielle*. Translated by Olivier Mannoni. Paris: Hachette, 2005.

Sloterdijk, Peter. *Spheres Volume 3: Foams: Plural Spherology*. Translated by Wieland Hoban. South Pasadena: Semiotext(e), 2016.

Smith, Daniel, and John Protevi. "Gilles Deleuze." In *The Stanford Encyclopedia of Philosophy*, edited by Edward N. Zalta (Winter 2015). Available online: https://plato.stanfod.edu/archives/win2015/entries/deleuze/

Sobchack, Vivian. "Between a Rock and a Hard Place: How Ridley Scott's *Prometheus* Deals with Impossible Expectations and Mythological Baggage." *Film Comment* (2012): 30–34.

Steadman, Phil. *Vermeer's Camera: Uncovering the Truth behind the Masterpieces*. Oxford: Oxford University Press, 2002.

Steinberg, Leo. "Other Criteria." In *Other Criteria: Confrontation with Twentieth-Century Art*. New York: Oxford University Press, 1972.

Styhre, Alexander. "The Concept of Transduction and Its Use in Organization Studies." *E:CO* 12, no. 3 (2010): 115–31.

Tomas, David. *Vertov, Snow, Farocki: Machine Vision and the Posthuman*. New York: Bloomsbury, 2013.

Väliaho, Pasi. *Biopolitical Screens: Image, Power, and the Neoliberal Brain*. Cambridge: The MIT Press, 2014.

Verne, Jules. *L'Île à hélice*. Paris: Union Générale d'Éditions, 1978.

Verne, Jules. *Propeller Island*. London: Granada Publishing, 1977.

Villaseñor, Maria-Christina. "Spectral Projections: Color, Race, and Abstraction in the Moving Image." In *Abstract Video: The Moving Image in Contemporary Art*, edited by Kate Mondloch and Gabrielle Jennings, 192–204. Oakland: University of California Press, 2015.

Virilio, Paul. *Bunker Archeology*. New York: Princeton Architectural Press, 1994.

Virilio, Paul. *Negative Horizon: An Essay in Dromoscopy*. Translated by Michael Degener. London: Continuum, 2007.

Vora, Kalindi. *Life Support: Biocapital and the New History of Outsourced Labor*. Minneapolis: University of Minnesota Press, 2015.

Welsh, Robert. "Molinari and the Science of Colour and Line." *RACAR* 1 (1978): 3–20.

Welu, James A. "Vermeer: His Cartographic Sources." *The Art Bulletin* 57, no. 4 (1975): 529–47.

Werner, Stephen. *Blueprint: A Study of Diderot and the Encyclopédie Plates*. Birmingham: Summa Publications, 1993.

Willatt, Edward. *Kant, Deleuze and Architectonics*. London: Continuum, 2010.

Williams, James. *Gilles Deleuze's Difference and Repetition: A Critical Introduction and Guide*. Edinburg: Edinburgh University Press, 2013.

Winkler, Harmut. "Computer and the Audiovisual: Ruptures and Continuities in Media History." In *ImageScapes: Studies in Intermediality*, edited by Christian J. Emden and Grabriele Rippl, 215–44. Oxford: Peter Lang, 2010.

Yang, Xin-She, and Xingshi He. "Swarm Intelligence and Evolutionary Computation: Overview and Analysis." In *Recent Advances in Swarm Intelligence and Evolutionary Computation*, edited by X.-S. Yang, 1–23. Cham: Springer, 2015.

Young, Eugene B. "Matter." In *The Deleuze and Guattari Dictionary*, edited by Eugene B. Young, Gary Genosko, and Janell Watson, 136–37. London: Bloomsbury, 2013.

Young, Eugene B., Gary Genosko, and Janell Watson, eds. *The Deleuze and Guattari Dictionary*. London: Bloomsbury Academic, 2013.

Zdebik, Jakub. *Deleuze and the Diagram: Aesthetic Threads in Visual Organization*. London: Continuum, 2012.

Zdebik, Jakub. "Skin Aesthetics as Incarnation: Gilles Deleuze's Diagram of Francis Bacon." *ESC* 34, no. 1 (March 2008): 149–164.

Zepke, Stephen. *Art as Abstract Machine: Ontology and Aesthetics in Deleuze and Guattari*. New York: Routledge, 2005.

Index

Note: Locators followed by 'n' denotes notes in the text respectively.